Schrödinger Equation - Fundamentals Aspects and Potential Applications

Edited by Muhammad Bilal Tahir, Muhammad Sagir, Muhammad Isa Khan and Muhammad Rafique

Published in London, United Kingdom

Schrödinger Equation - Fundamentals Aspects and Potential Applications
http://dx.doi.org/10.5772/intechopen.111032
Edited by Muhammad Bilal Tahir, Muhammad Sagir, Muhammad Isa Khan and Muhammad Rafique

Contributors
Hocine Boukabcha, Aid Salah Eddine, Ghobrini Amina, Noorul Ain, Sadaf Fatima, Mushtaq Ahmad, Muhammad Aslam, Muhammad Rizwan Khan, Gabino Torres-Vega, Armando Martinez-Perez, José Juan Peña, Jesús Morales, Jesús García-Ravelo, Daniel R. E. Hodgson, Juan Antonio Nieto, George Vahala, Min Soe, Kyriakos Hitzanidis, Linda Vahala, Abhay K. Ram, Efstratios Koukoutsis, Masaru Hamano, Masahiro Ikeda

First published in London, United Kingdom, 2024 by IntechOpen
IntechOpen is the global imprint of INTECHOPEN LIMITED, registered in England and Wales, registration number: 11086078, 5 Princes Gate Court, London, SW7 2QJ, United Kingdom

British Library Cataloguing-in-Publication Data
A catalogue record for this book is available from the British Library

Additional hard and PDF copies can be obtained from orders@intechopen.com

Schrödinger Equation - Fundamentals Aspects and Potential Applications
Edited by Muhammad Bilal Tahir, Muhammad Sagir, Muhammad Isa Khan and Muhammad Rafique
p. cm.
Print ISBN 978-1-83769-213-2
Online ISBN 978-1-83769-214-9
eBook (PDF) ISBN 978-1-83769-215-6

For EU product safety concerns:
IN TECH d.o.o., Prolaz Marije Krucifikse Kozulić 3, 51000 Rijeka, Croatia,
info@intechopen.com or visit our website at intechopen.com.

Meet the editors

Dr. Muhammad Bilal Tahir is a distinguished physicist and academic leader who currently serves as the director of the Institute of Physics. With a career dedicated to the advancement of physics and education, he has made significant contributions to the field. He holds a Ph.D. in Physics, earned through rigorous research and a commitment to excellence. Dr. Tahir's expertise spans several areas of physics, including quantum mechanics, condensed matter physics, and theoretical physics, making him a versatile and well-rounded scientist. As the director of the Institute of Physics, Dr. Tahir has played a pivotal role in shaping the institution's academic and research programs. His visionary leadership has driven the development of cutting-edge research initiatives and cultivated a vibrant academic community. Under his guidance, the institute has flourished as a center of excellence for both undergraduate and graduate students, fostering a spirit of innovation and inquiry. Beyond his administrative role, Dr. Tahir remains an active researcher, with numerous publications in reputable scientific journals. His work reflects a deep commitment to advancing our understanding of the fundamental principles that govern the physical world. Dr. Tahir's dedication to physics and education has left an indelible mark on the academic community and the broader scientific landscape. His passion for research, teaching, and leadership continues to inspire the next generation of physicists and scholars.

Professor Dr. Muhammad Sagir is a highly accomplished academic and professional in chemical and environmental engineering. Currently serving as the director of the Institute of Chemical and Environmental Engineering, Khawaja Fareed University of Engineering and Information Technology (KFUEIT), Pakistan, his career has been marked by exceptional contributions to the field. His passion for teaching, research, and academic administration continues to inspire students, researchers, and professionals alike.

Dr. Muhammad Isa Khan is an esteemed academic and scholar who is currently an Associate Professor in Physics at The Islamia University of Bahawalpur, Pakistan. With a career dedicated to education and research, Dr. Khan has made significant contributions to his field. Dr. Khan holds a Ph.D. in Physics, a testament to his dedication to academic excellence and a profound understanding of his subject matter. His expertise spans several domains, reflecting his versatility and comprehensive knowledge in his area of expertise. Dr. Khan is an active researcher with numerous publications in prestigious academic journals. His work reflects a deep commitment to advancing knowledge in his field, and his research has made a tangible impact on the academic and professional community.

Dr. Muhamamd Rafique holds a Ph.D. in Physics, which serves as a testament to his dedication to academic rigor and the depth of his subject knowledge. His expertise spans various domains, and he has established himself as a versatile and well-rounded academic. In his role as an assistant professor at the University of Sahiwal, Pakistan, Dr. Rafique has made substantial contributions to the institution's academic and research programs. His commitment to teaching and mentoring students has helped foster a vibrant academic community, and his guidance has proven invaluable in shaping the academic journeys of many aspiring scholars.

Contents

Preface

The 20th century bore witness to two monumental scientific revolutions that forever altered the landscape of our understanding of the universe. First came the theory of relativity, which fundamentally transformed our conceptions of space and time, weaving a tapestry of spacetime that reshaped the way we perceive the fabric of the cosmos. Concurrently, quantum mechanics emerged, a groundbreaking framework that revolutionized our comprehension of the states of matter and energy, delving into the depths of the microcosmic world. Quantum mechanics, with its enigmatic principles and mathematical intricacies, has emerged as a captivating and challenging subject in the realms of science, engineering, and technology. It serves as the key to unlocking a wide array of phenomena, unveiling the beauty of mathematical techniques, Schrödinger's equation, and the ethereal realm of wave functions. It has enabled us to probe the mysteries of matter and energy at their most fundamental levels.

It is with great enthusiasm that we present this book, *Schrödinger Equation – Fundamental Aspects and Potential Applications*. Within its pages, we explore a myriad of topics encompassing the essence of quantum mechanics. Our journey commences with an introduction to the spectral theory of the one and multi-dimensional Schrödinger equation, offering a foundation for understanding its mathematical underpinnings. As we delve deeper, we venture into the realm of scattering theory for the multi-dimensional non-relativistic Schrödinger equation, shedding light on the intricacies of quantum interactions. The quantization process and the profound concept of Feynman path integrals also find their place within our exploration as we provide insight into the mechanics of quantum systems and their behavior. Moreover, we examine the innovative application of relative entropy methods and transformation theory, unveiling their significance in constructing singular diffusion processes that bear an astonishing resemblance to Schrödinger equations. These methodologies provide fresh perspectives on the underlying principles governing quantum dynamics.

We have endeavored to make this book accessible, well-organized, and self-contained, ensuring that it caters to a broad audience, including undergraduates, graduates, researchers, and professionals in the fields of functional analysis, probability theory, and quantum dynamics. Our aim is to facilitate a comprehensive understanding of the Schrödinger equation and its multifaceted applications. We trust that readers will find this book to be a valuable resource, brimming with profound insights into the Schrödinger equation and its far-reaching implications. This book is a comprehensive guide that both educates and inspires, fostering a deeper appreciation for the profound wonders of quantum mechanics.

We invite you to embark on this journey with us, as we unravel the mysteries of the Schrödinger equation and explore the boundless potential it holds in the realms of science, engineering, and technology.

Welcome to *Schrödinger Equation – Fundamental Aspects and Potential Applications*.

Muhammad Bilal Tahir
Director,
Institute of Physics,
Khwaja Fareed University of Engineering and Information Technology,
Rahim Yar Khan, Pakistan

Muhammad Sagir
Director,
Institute of Chemical and Environmental Engineering,
Khawaja Fareed University of Engineering and Information Technology (KFUEIT),
Rahim Yar Khan, Pakistan

Muhammad Isa Khan
Associate Professor,
Islamia University of Bahawalpur,
Bahawalpur, Pakistan

Muhamamd Rafique
Assistant Professor,
University of Sahiwal,
Sahiwal, Pakistan

Section 1

Schrödinger Equation – Fundamentals Aspects

Chapter 1

Schrödinger Wave Equation for Simple Harmonic Oscillator

Noor-ul-ain, Sadaf Fatima, Mushtaq Ahmad, Muhammad Rizwan Khan and Muhammad Aslam

Abstract

In physics, harmonic motion is among the most representative types of motion. A simple harmonic oscillator is often the source of any vibration with a restoring force proportional to Hooke's law. Every minimum potential has a solution in the form of the harmonic oscillator potential. Little oscillations at the minimum are characteristic of almost all natural potentials and of many quanta mechanical systems. Harmonic motion is an essential building block for these more complex uses. The Schrödinger equation is a defining feature of the harmonic oscillator. Here, we demonstrate that the time-frequency plane is a useful tool for analyzing their dynamics. We numerically integrate several examples involving different input forces and demonstrate that the oscillations are clearly displayed and easily interpretable in the time-frequency plane.

Keywords: harmonic motion, frequency, pendulum, displacement, amplitude

1. Introduction

A system that uses simple harmonic motion (SHM) is known as a harmonic oscillator.

A physical system called a harmonic oscillator experiences a restoring force proportionate to the displacement when it is moved away from its mean position.

A wave equation that describes the behavior of quantum particles is the Schrödinger equation. A harmonic oscillator's energy levels can be demonstrated by the Schrödinger equation to be quantized, which means that they can only take on specific discrete values. The Schrödinger equation has the effect of restricting the possible energies that an oscillator that is harmonic can have [1, 2].

A physical system known as harmonic oscillator oscillates at a frequency proportional to the displacement from its equilibrium position and is governed by a restoring force F_r. The F_r is proportional to the displacement from its mean position. This means that the system tends to return to its equilibrium position when disturbed from it, and the rate at which it oscillates is determined by the strength of the restoring force and the mass of the system. An equation of simple harmonic motion which is sinusoidal function of time with constant amplitude and frequency can be used to describe the motion of harmonic oscillator [1, 3]. The two examples of harmonic oscillator are mass connected to the spring and a simple pendulum. Harmonic oscillators are important in physics and engineering because they provide a useful model for many physical systems and can be used to analyze and predict the behavior of those systems [3, 4].

2. Classical behavior of simple harmonic oscillator

The simple example of linear harmonic oscillator is a mass attached to a wall by means of a spring as illustrated in the following **Figure 1**.

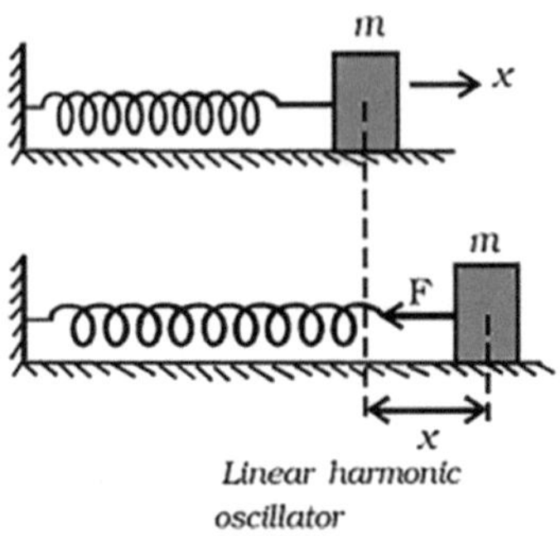

Figure 1.
Shows the experimental device for the study of the spring-mass system [1].

2.1 Expression for potential energy of simple harmonic oscillator

Hooke's law states that the force required to stretch or compress a spring is proportionate to the distance extended or compressed from its original length. Mathematically, this relation can be expressed as:

$$F \propto x$$

$$F \propto -x$$

$$F_r = -kx \tag{1}$$

Where, F_r is the force applied to the spring, x is the displacement of the spring from the original length, and k is a constant which is known as spring constant and represents the stiffness of spring [5].

Hooke's law applies to all elastic materials, not just springs. It is an important concept in physics and engineering because it helps to understand and predict the behavior of systems that involve elastic materials, such as springs, rubber bands, and other materials. Hooke's law is also the basis for the design of many mechanical systems, such as shock absorbers, suspension systems, and other devices that rely on the properties of elastic materials [6, 7].

When an object is displaced from its equilibrium position, a restoring force acts on it to push or pull it back toward that position. The F_r is directly proportional to the displacement from the equilibrium position and also acts in opposite direction [5]. This force is present in many physical systems, such as springs, pendulums, and mass-spring systems, and it plays a vital role in the behavior of these systems [3, 4].

$$F = -\frac{dv}{dx} \tag{2}$$

∵Force F can be expressed as negative derivative of potential energy V.
The work done in stretching spring to distance dx

$$W = F \times \text{distance}$$
$$P.E = F \times dx$$
$$-dv = F \times dx$$
$$dv = -F \times dx$$
$$dv = -F \times dx$$

From Eq. (1)

$$F = -kx$$
$$dv = kx \times dx \tag{3}$$

Integrate Eq. (3) within limits 0 → x

$$\int dv = +\int_0^x kxdx$$
$$V = k\int_0^x xdx$$
$$V = k \lim_{0 \to x} \frac{x^2}{2}$$
$$V = k\left(\frac{x^2}{2} - \frac{0}{2}\right)$$
$$V = k\frac{x^2}{2}$$
$$V = \frac{1}{2}kx^2 \tag{4}$$

Where x is the distance from equilibrium position [8, 9].

The plot of potential energy (V) of a particle executing simple harmonic motion against displacement from its equilibrium length is a parabola as illustrated in the following **Figure 2**.

2.2 Expression for frequency of linear harmonic oscillator

The frequency of a harmonic oscillator is the number of complete oscillations or cycles it completes per unit time. The frequency of a harmonic oscillator depends on the physical characteristics of the system, such as its mass and stiffness.

According to second law of motion

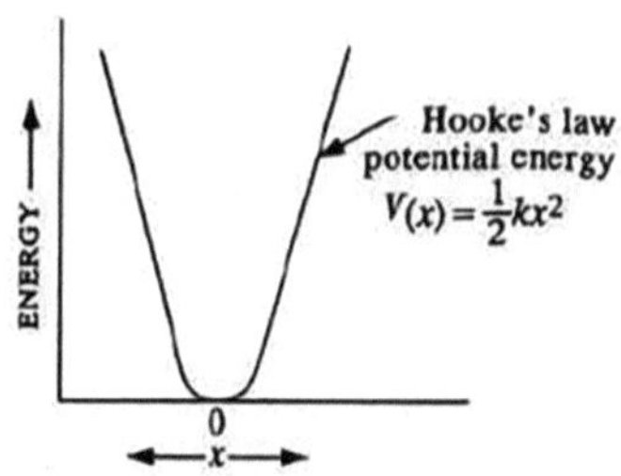

Figure 2.
The potential energy for a simple harmonic oscillator [6].

$$\mathrm{F = ma} \tag{5}$$

Comparing Eqs. (1) and (5)

$$\mathrm{ma = -kx}$$

$$m\frac{d^2x}{dt^2} = -kx \because a = \frac{d^2x}{dt^2}$$

$$\frac{d^2x}{dt^2} + \frac{k}{m}x = 0 \tag{6}$$

Eq. (6) is a second-order differential equation. The general solution of this Eq. (6)

$$x = A\sin\omega t \tag{7}$$

We know $\omega = \sqrt{\frac{k}{m}}$

$$x = A\sin\sqrt{\frac{k}{m}}t \tag{8}$$

We know that

$$\omega = 2\pi\vartheta t$$

$$x = A\sin 2\pi\vartheta \tag{9}$$

Comparing Eqs. (8) and (9)

$$A\sin\sqrt{\frac{k}{m}}t = A\sin 2\pi\vartheta t$$

$$\sin\sqrt{\frac{k}{m}} = \sin 2\pi\vartheta$$

$$\sin^{-1}\sin\sqrt{\frac{k}{m}} = \sin^{-1}\sin 2\pi\vartheta$$

$$\because\sqrt{\frac{k}{m}} = 2\pi\vartheta$$

$$\vartheta = \frac{1}{2\pi}\sqrt{\frac{k}{m}} \tag{10}$$

Eq. (10) gives the frequency of the simple harmonic oscillator, where ϑ the frequency, k is the spring constant, and m is the mass of a linear harmonic oscillator. The above equation determines that the frequency of a harmonic oscillator is directly proportional to spring constant's square root and inversely proportional to mass's square root. This means that by increasing the stiffness of the spring or by decreasing the mass of the oscillator, the frequency of an oscillator will increase [8].

Generally, the frequency of harmonic oscillator is an important characteristic that determines its behavior and can be used to analyze and predict its motion. The frequency of a harmonic oscillator can be measured experimentally using various methods, such as by measuring the time period of its oscillations or by analyzing its response to external forces.

$$\because \vartheta = \frac{c}{\lambda} \nabla = \frac{1}{\lambda} \vartheta = c\nabla$$

$$2\pi c\nabla = \sqrt{\frac{k}{m}}$$

$$\nabla = \frac{1}{2\pi c}\sqrt{\frac{k}{m}}$$

∇ is wave number

For two particles connected to each other through a spring as in diatomic molecule, we use term reduced mass μ [10].

$$\nabla = \frac{1}{2\pi c}\sqrt{\frac{k}{\mu}} \tag{11}$$

3. Quantum mechanical treatment of simple harmonic oscillator

The wave function is a mathematical representation of a quantum system's state in quantum mechanics. All of the information about a particle or a group of particles, including their position, momentum, and energy, is contained in the wave function. It is a complex-valued function depends on position and time of particle. It is denoted by symbol Ψ [11].

Probability of finding the particle at a certain position is proportional to absolute square of wave function. It is also used to determine the probability density of finding a particle within a certain volume of space.

In quantum mechanics, wave function is a fundamental concept used to calculate many properties of quantum systems, such as energy levels, transition probabilities, and scattering cross-sections. The wave function is also used to describe the behavior of systems that exhibit wave-like properties, such as electrons, atoms, and molecules [12, 13].

The wave function follows the Schrödinger equation, which is a differential equation that determines how the wave function evolves over time. The Schrödinger equation is used to determine the temporal evolution of quantum systems and to predict particle and system behavior under different conditions [14].

3.1 Representation of wave function

In quantum mechanics, the wave function can be represented in several ways, depending on the context and the physical system being described. Here are three common representations [15]:

3.1.1 Position representation

In this position representation, $\Psi(x,t)$ gives the probability amplitude of finding a particle at position x at time t. The position representation is used for systems with definite position, such as single particle in a box or a molecule. In this representation, a wave function is typically denoted as $\Psi(x,t)$ i.e., function of position and time. Its mathematical form can be written as: $\Psi(x,t) = A(x,t) * \exp(i\varphi(x,t))$ where $A(x,t)$ is the amplitude of the wave function and $\varphi(x,t)$ is its phase. The amplitude is a real-valued function that describes the intensity of the wave, while the phase is a real-valued function that describes the position of the wave in space and time [16, 17].

3.1.2 Momentum representation

In this representation, the wave function is function of momentum rather than the position. The wave function $\Psi(p,t)$ gives the probability amplitude of finding a particle with momentum p at time t. The momentum representation is useful for systems with definite momentum, like a free particle. In this representation, wave function is typically denoted as $\Psi(p,t)$ and is function of momentum and time. Its mathematical form can be written as: $\Psi(p,t) = B(p,t) * \exp(i\chi(p,t))$ where $B(p,t)$ is the amplitude of the wave function in momentum space and $\chi(p,t)$ is its phase. This amplitude is real-valued function that determines the intensity of the wave in momentum space, while the phase is a real-valued function that describes the position of wave in momentum space [16–18].

3.1.3 Energy representation

In this representation, the wave function is a function of energy. The wave function $\Psi(E)$ gives the probability amplitude of finding a system with energy E. The energy representation is useful for systems with definite energy, like a particle in the potential well. In the energy representation, wave function is typically denoted as $\Psi(E)$ and is a function of energy. Mathematically, it can be written as

$$\Psi(E) = C(E) * \exp(i\psi(E)) \tag{12}$$

where $C(E)$ is the amplitude of the wave function in energy space and $\psi(E)$ is its phase. This amplitude is real-valued function that determines the intensity of wave in energy space, while the phase is a real-valued function that describes the position of the wave in energy space.

In each representation, Ψ is a complex-valued function satisfies the Schrödinger equation. It can be normalized, which means that the integral of the absolute square of the wave function over all space or momentum or energy is equal to one, ensuring that the probability of locating a particle in the system is one.

The mathematical form of wave function can be used to calculate various properties of the system, such as probabilities of finding the particle in a certain position, momentum, or energy state [19].

3.2 Boundaries conditions

For the harmonic oscillator, the two common boundary conditions are described as follows [20].

3.2.1 Normalizability condition

The wave function must be normalizable, which means that the integral of the absolute square of the wave function over all space must be finite. This assures that probability of locating a particle in the system is one [19].

3.2.2 Continuity condition

The wave function must be continuous and differentiable at the ends of the range. This ensures that the probability density and its first derivative are continuous and smooth throughout the range of motion.

For the harmonic oscillator, the boundary conditions are typically satisfied by using a particular type of wave function, called the Hermite polynomials. The Hermite polynomials are a set of orthogonal polynomials that satisfy both the normalizability and continuity conditions. They form a complete basis set for the wave function of the harmonic oscillator, allowing the solution to be expressed as a linear combination of these polynomials [6].

3.3 Schrödinger wave equation for harmonic oscillator

The mathematical form of the wave function in quantum mechanics depends on the physical system being described and the representation being used. However, in general, it is a complex-valued function that satisfies Schrödinger equation [8, 21].

In Quantum mechanics, the one-dimensional time-independent Schrödinger wave equation for harmonic oscillator follows as [22]:

$$\frac{\partial^2 \psi}{\partial x^2} + \frac{2m}{\hbar^2}(E - V)\psi = 0 \tag{13}$$

But the potential energy of the simple harmonic oscillator is $V=\frac{1}{2}Kx^2$, therefore

$$\frac{\partial^2 \psi}{\partial x^2} + \frac{2m}{\hbar^2}\left(E - \frac{1}{2}Kx^2\right)\psi = 0 \tag{14}$$

Or

$$\frac{\partial^2 \psi}{\partial x^2} + \frac{-mKx^2}{2}\psi = \frac{-2mE}{2}\psi$$

$$\frac{mK}{\hbar^2} = \alpha^2 \frac{2mE}{\hbar^2} = \varepsilon$$

Them

$$\frac{\partial^2\psi}{\partial x^2} - \alpha^2 x^2\, \psi = -\varepsilon\psi \tag{15}$$

This is Schrödinger's equation for harmonic oscillator [23–25]. Here x^2 is the coefficient of ψ, so it is difficult to obtain its solution. Hence we will find its asymptotic solution

When $x \rightarrow \infty\ \alpha^2 x^2 > > \varepsilon$

So we can write:

$$\frac{\partial^2\psi}{\partial x^2} - \alpha^2 x^2\, \psi = 0 \tag{16}$$

Its solution is $\psi = e^{\pm\alpha x^2/2}$

$$\frac{\partial\psi}{\sigma x} = \pm\alpha x e^{\pm\alpha x^2/2}$$

$$\frac{\partial^2\psi}{\partial x^2} = \frac{\partial}{\sigma x}\left(\pm\alpha x e^{\pm\alpha x^2/2}\right) = \alpha^2 x^2 e^{\pm\alpha x^2/2} \pm \alpha e^{\pm\alpha x^2/2} = (\pm\alpha)e^{\pm\alpha x^2/2}\alpha^2 x^2$$

Value of αx is larger hence we take $(\alpha^2 x^2 \pm \alpha) \approx \alpha^2 x^2$

$$\frac{\partial^2\psi}{\partial x^2} = \alpha^2 x^2 e^{\pm\alpha x^2/2}$$

Or $\frac{\partial^2\psi}{\partial x^2} = \alpha^2 x^2\,\psi$ or $\frac{\partial^2\psi}{\partial x^2} - \alpha^2 x^2\,\psi = 0$

Now we take $\psi = e^{-\alpha x^2/2}$

Because it obeys the condition that $|\psi|^2$ decreases with increasing x

General solution:

$$\psi_{(x)=}f_{(x)}e^{-\alpha x^2/2}$$

Differentiating w.r.t x

$$\frac{\partial\psi}{\sigma x} = f_{(x)}e^{-\alpha x^2/2}(-\alpha x) + e^{-\alpha x^2/2}\frac{\partial f}{\partial x}$$

Again differentiating w.r.t x

$$\frac{\partial^2\psi}{\partial x^2} = f_{(x)}\left[e^{-\alpha x^2/2}(-\alpha) + (-\alpha x)(-\alpha x)e^{-\alpha x^2/2}\right]$$

$$+(-\alpha x)\, e^{-\alpha x^2/2}\frac{\partial f}{\partial x} + \frac{\partial f}{\partial x}e^{-\alpha x^2/2}\left(-\alpha x + e^{-\alpha x^2/2}\frac{\partial^2 f}{\partial x^2}\right.$$

$$\frac{\partial^2\psi}{\partial x^2} = e^{-\alpha x^2/2}f_{(x)}\left(-\alpha + \alpha^2 x^2\right) + \frac{\partial f}{\partial x}e^{-\alpha x^2/2}(-2\alpha x) + e^{-\alpha x^2/2}\frac{\partial^2 f}{\partial x^2}$$

$$\frac{\partial^2\psi}{\partial x^2} = e^{-\alpha x^2/2}\left[\frac{\partial^2 f}{\partial x^2} - 2\alpha x\,\frac{\partial f}{\partial x} + \left(\alpha^2 x^2 - \alpha\right)f\right]$$

Substituting values of ψ and $\frac{\partial^2\psi}{\partial x^2}$ in Eq. (15)

$$e^{-\alpha x^2/2}\left[\frac{\partial^2 f}{\partial x^2}-2\alpha x\,\frac{\partial f}{\partial x}+\left(\alpha^2x^2-\alpha\right)\mathrm{f}\right]-\alpha^2x^2fe^{-\alpha x^2/2}=-\varepsilon fe^{-\alpha x^2/2}$$

$$\text{Or }\frac{\partial^2 f}{\partial x^2}-2\alpha x\,\frac{\partial f}{\partial x}+(\varepsilon-\alpha)f=0 \tag{17}$$

Now substituting y = $\sqrt{\alpha}$ x and $f_{(x)}=H_{(y)}$ converting into standard Hermite polynomial equation

y = $\sqrt{\alpha}\,x$ then $\frac{dy}{dx}=\sqrt{\alpha}$

$$\frac{\partial f}{\partial x}=\frac{\partial f}{\partial y}\cdot\frac{\partial y}{\partial x}=\sqrt{\alpha}\frac{\partial f}{\partial y}$$

$$\frac{\partial^2 f}{\partial x^2}=\frac{\partial}{\partial x}\left(\frac{\partial f}{\partial x}\right)=\frac{\partial}{\partial x}\left(\sqrt{\alpha}\frac{\partial f}{\partial y}\right)=\frac{\partial}{\partial y}\left(\sqrt{\alpha}\frac{\partial f}{\partial y}\right)\frac{\partial y}{\partial x}=\alpha\frac{\partial^2 f}{\partial y^2}$$

Substituting values of $\frac{\partial f}{\partial x}$ and $\frac{\partial^2 f}{\partial x^2}$ in Eq. (17), we get

$$\alpha\frac{\partial^2 f}{\partial y^2}-2\alpha\frac{y}{\sqrt{\alpha}}\sqrt{\alpha}\frac{\partial f}{\partial y}+(\varepsilon-\alpha)f=0$$

$$\alpha\frac{\partial^2 f}{\partial y^2}-2\alpha y\,\frac{\partial f}{\partial y}+(\varepsilon-\alpha)f=0$$

$$\frac{\partial^2 f}{\partial y^2}-2y\,\frac{\partial f}{\partial y}+\left(\frac{\varepsilon}{\alpha}-1\right)f=0$$

Now f(x) = H(y)

$$\frac{\partial^2 H}{\partial y^2}-2y\,\frac{\partial H}{\partial y}+\left(\frac{\varepsilon}{\alpha}-1\right)H=0 \tag{18}$$

This is standard Hermite differential equation [22]. It can be expressed as

$$\mathrm{H(y)}=\sum_{p=0}^{\infty}a_py^p \tag{19}$$

$$\frac{\partial H}{\partial y}=\sum pa_py^{p-1}$$

$$\frac{\partial^2 H}{\partial y^2}=\sum p(p-1)a_py^{p-2}$$

From Eq. (18)

$$\sum p(p-1)a_py^{p-2}-\sum\left[2p-\left(\frac{\varepsilon}{\alpha}-1\right)\right]a_py^p=0$$

This expression is valid only when coefficient of each power of *y* is zero.
And p = p + 2

$$\sum(p+2)(p+2-1)a_{p+2}y^{p+2-2} - \sum\left[2p - \left(\frac{\varepsilon}{\alpha} - 1\right)\right]a_p y^p = 0$$

$$a_{p+2}(p+2)(p+1) = a_p\left[2p - \left(\frac{\varepsilon}{\alpha}\right) + 1\right]$$

$$a_{p+2} = \frac{\left[2p - \left(\frac{\varepsilon}{\alpha}\right) + 1\right]}{(p+2)(p+1)}a_p \tag{20}$$

We can determine values of all the coefficients in terms of two arbitrary constants a_0 and a_1

Thus, complete solution of Schrödinger's equation is [26]

$$\psi = e^{-\alpha x^2/2}H_{(y)}$$

$$\psi = e^{-y^2/2}H_{(y)}$$

3.4 Energy eigen values

$\psi = e^{-y^2/2}H_{(y)}$ of a simple harmonic oscillator will be physically accepted only when y→ ∞, the increase in the value of Hermite Polynomial $H_{(y)}$ is more rapid than the decrease in the value of $e^{-y^2/2}$ value [27].

Value of $e^{-y^2/2}H_{(y)}$ can be zero only when power series for $H_{(y)}$ is finite series.

Let series be finite for p=n, the Eq. (20) becomes.

$$2\text{n} - \frac{\varepsilon}{\alpha} + 1 = 0$$

$$\text{N} = \frac{1}{2}\left(\frac{\varepsilon}{\alpha} - 1\right)\frac{\varepsilon}{\alpha} = 2n + 1\varepsilon = \frac{2mE}{\hbar^2}\alpha = \sqrt{\frac{mk}{\hbar^2}}$$

$$\frac{2mE}{\hbar^2}\Big/\sqrt{\frac{mk}{\hbar^2}} = 2\text{n} + 1\,\frac{2}{\hbar}\sqrt{\frac{m}{k}}E = 2n + 1$$

$$\text{E} = \frac{2n+1}{2}\hbar\sqrt{\frac{k}{m}}$$

But we know $\sqrt{\frac{k}{m}} = \omega$ (angular frequency)

$$\text{E} = \left(\frac{2n+1}{2}\right)\hbar\omega = \left(n + \frac{1}{2}\right)\hbar\nu \tag{21}$$

Where n = 0, 1, 2, 3, ...

The above equation gives the energy levels of a harmonic oscillator [28], where n is a non-negative integer, *h* is reduced Planck constant, *ω* is an angular frequency of the oscillator, and E_n is the energy of the oscillator in the nth energy level. In quantum mechanics, the energy levels of simple harmonic oscillator are quantized, which means they take on only certain discrete values.

If n = 0 then $E_0 = \frac{1}{2}\,\nu$
n = 1 then $E_1 = \frac{3}{2}\,\nu$
n = 2 then $E_2 = \frac{5}{2}\,\nu$

The energy levels of a harmonic oscillator are equally spaced, with the energy of each level separated by an amount $h\ \omega$. The ground state of the oscillator, n=0, has the lowest energy level and corresponds to the oscillator's minimum energy state, where the particle is localized at the center of the potential well. As n increases, the energy levels increase and the wave function oscillates with more nodes [27].

The energy of the harmonic oscillator is always positive, and the oscillator can never reach the zero-point energy, which is the minimum possible energy that a quantum mechanical system can have [29].

Author details

Noor-ul-ain[1*], Sadaf Fatima[1], Mushtaq Ahmad[1], Muhammad Rizwan Khan[1] and Muhammad Aslam[2]

1 Institute of Physics, The Islamia University of Bahawalpur, Pakistan

2 Institute of Physics and Technology, Ural Federal University, Yekaterinburg, Russia

*Address all correspondence to: noorulain@iub.edu.pk

References

[1] Triana C, Fajardo F. The influence of spring length on the physical parameters of simple harmonic motion. European journal of physics. 2011;**33**(1):219

[2] Garrett SL. Understanding Acoustics: An Experimentalist's View of Sound and Vibration. Springer Nature; 2020

[3] Garrett SL. Understanding Acoustics. In: Pine Grove Mills. PA, USA: springer; 2020. pp. 59-131

[4] Cheney CP. Simple Pendulum and Hooke's Law Prelab. 2018

[5] Clark D, Franklin J, Mann N. Relativistic linear restoring force. European Journal of Physics. 2012;**33**(5): 1041

[6] Guiard Y. On Fitts's and Hooke's laws: Simple harmonic movement in upper-limb cyclical aiming. Acta Psychologica. 1993;**82**(1-3):139-159

[7] Struganova I. A spring, Hooke's law, and Archimedes' principle. The Physics Teacher. 2005;**43**(8):516-518

[8] Schrodinger Wave Equation for a Linear Harmonic Oscillator & Its Solution by Polynomial Method

[9] Dantas CM, Pedrosa I, Baseia B. Harmonic oscillator with time-dependent mass and frequency and a perturbative potential. Physical Review A. 1992;**45**(3):1320

[10] Wierling A, Sawada I. Wave-number dependent current correlation for a harmonic oscillator. Physical Review E. 2010;**82**(5):051107

[11] Song D-Y. Unitary relation between a harmonic oscillator of time-dependent frequency and a simple harmonic oscillator with and without an inverse-square potential. Physical Review A. 2000;**62**(1):014103

[12] Dekker H. Classical and quantum mechanics of the damped harmonic oscillator. Physics Reports. 1981;**80**(1): 1-110

[13] Senitzky I. Dissipation in quantum mechanics. The harmonic oscillator. Physical Review. 1960;**119**(2):670

[14] Guedes I. Solution of the Schrödinger equation for the time-dependent linear potential. Physical Review A. 2001;**63**(3):034102

[15] Oh H et al. Exact wave functions and coherent states of a damped driven harmonic oscillator. Physical Review A. 1989;**39**(11):5515

[16] Dahl JP, Springborg M. The Morse oscillator in position space, momentum space, and phase space. The Journal of chemical physics. 1988;**88**(7):4535-4547

[17] Robinett R. Quantum and classical probability distributions for position and momentum. American Journal of Physics. 1995;**63**(9):823-832

[18] Rushka M, Freericks J. A completely algebraic solution of the simple harmonic oscillator. American Journal of Physics. 2020;**88**(11):976-985

[19] Swenson R, Hermanson J. Energy quantization and the simple harmonic oscillator. American Journal of Physics. 1972;**40**(9):1258-1260

[20] Zhang K et al. Simple harmonic oscillation in a non-Hermitian Su-Schrieffer-Heeger chain at the exceptional point. Physical Review A. 2018;**98**(2):022128

[21] Davey ACH. Schrodinger wave equation. MacEwan University Student eJournal. 2020;**4**(1)

[22] Amir N, Iqbal S. Exact solutions of Schrödinger equation for the position-dependent effective mass harmonic oscillator. Communications in Theoretical Physics. 2014;**62**(6):790

[23] Chow P. Computer solutions to the Schrödinger equation. American Journal of Physics. 1972;**40**(5):730-734

[24] Mills K, Spanner M, Tamblyn I. Deep learning and the Schrödinger equation. Physical Review A. 2017; **96**(4):042113

[25] University of Northern Iowa. The Schröedinger Equation in One Dimension. Available from: https://faculty.chas.uni.edu/~shand/Mod_Phys_Lecture_Notes/Chap7_Schrodinger_Equation_1D_Notes_s12.pdf

[26] Taşeli H. On the exact solution of the Schrödinger equation with a quartic anharmonicity. International Journal of Quantum Chemistry. 1996;**57**(1):63-71

[27] Ram-Mohan LR et al. The finite-element method for energy eigenvalues of quantum mechanical systems. Computers in Physics. 1990;**4**(1):50-59

[28] Biswas S et al. Eigenvalues of λ x2m anharmonic oscillators. Journal of Mathematical Physics. 1973;**14**(9):1190-1195

[29] Fernández FM. On the singular harmonic oscillator. arXiv preprint arXiv:2112.03693, 2021

Chapter 2

A Schrödinger Equation for Light

Daniel R.E. Hodgson

Abstract

In this chapter we examine the quantised electromagnetic (EM) field in the context of a Schrödinger equation for single photons. For clarity we consider only a one-dimensional system. As a universal tool for calculating the time-evolution of quantum states, a Schrödinger equation must exist that describes the propagation of single photons. Being inherently relativistic, however, critical aspects of both special relativity and quantum mechanics must be combined when quantising the EM field. By taking the approach of a Schrödinger equation for localised photons, we will show how novel and previously overlooked features of the quantised EM field become a necessary part of a complete description of photon dynamics. In this chapter, I shall provide a thorough examination of new features and discuss their significance in topics such as quantum relativity and photon localisation.

Keywords: photon localisation, photon wave function, causality, negative frequencies, non-locality

1. Introduction

Thomas Young's double slit experiment gives a simple but clear demonstration that light is certainly a wave. The appearance of an alternating pattern of dark fringes is evidence of the destructive superposition of waves passing through different slits onto the screen behind. In this classical experiment, the pattern emerging on the screen is generated by the interference between oscillating electromagnetic (EM) waves that are predicted by Maxwell's theory of electromagnetism. The modification from a classical to a quantum theory, however, reinterprets these waves as oscillations of the probabilistic wave function for a collection of photons, the indivisible particles of light. In order to form a complete description of how photons evolve, it is important that we are able to define a wave function for each photon wave packet describing its oscillations through both space and time.

By initially postulating that photons are discrete and countable objects, and that each photon has an energy proportional to its frequency, it is possible to derive complete expressions for the electric and magnetic field observables up to an overall phase [1]. More conventional quantisation methods, however, take the reverse approach. Here the field observables, not the photons, are the main focus of the quantisation process, which are obtained by means of canonical quantisation. See, for example, Ref. [2]. By adopting a Hamiltonian procedure, correspondence with classical physics can be maintained by imposing canonical commutation relations. Moreover, working directly with quantised fields may be viewed as more fundamental than working with particles, which are not covariant objects. Providing a wave function for the excitations of the field observables, however, has proved exceptionally challenging.

It was proven in 1948 by Newton and Wigner that no position-dependent wave function existed for the photon [3, 4]. More specifically, subject to certain conditions, there was no photon position operator with which to define a basis of localised eigenstates for the wave function. Since this time, the localisation of single photons has been researched extensively [5–12], but there is as yet no unanimous agreement on whether localisation is possible. In the work of Fleming [13, 14], for example, some of the relativistic properties of the Newton-Wigner (NW) operator were clarified, but localisation of the photon was again shown to be impossible. Only more recently, by considering the longitudinal components of spin, Hawton has been able to show that a photon position operator with commuting components conjugate to the momentum operator can be defined [15–17].

It is also unclear how the photon wave function ought to be interpreted. Early wave functions, such as the Landau-Peierls wave function [18–20], for example, were criticised for being non-locally related to the electric and magnetic field observables. Such a relationship emerges due to a disparity between the units for a probabilistic wave function and the field observables. Wave functions locally related to the field observables have been studied in both the first and second quantisation regimes [21–25]. In some schemes, such as those of Knight [26] and Licht [27, 28], a photon is only localised if the electric and magnetic field expectation values are also localised. In this case, however, a single photon cannot be localised if the field observables do not commute [29]. To overcome problems of non-locality, many authors have introduced non-local inner products, which lead to the use of non-standard and non-Hermitian models [30–33].

Localisation in quantum theory is also closely connected to causality. A photon wave packet that is localised to one region, for instance, cannot reach another until a time has elapsed no less than the distance between these regions divided by the speed of light in a vacuum. The theorems of Hegerfeldt [34] and Malament [35] show, however, that non-zero correlations between the position of a wave packet can be generated at speeds exceeding the speed of light. The question that this raises about causality has been a large topic of research [36–40], with a particular interest in causality in the transmission of radiation between two two-level atoms [41–44]. Whilst many insist that only causality in the sense of no-signalling, rather than of strict causality, is necessary, a wave function can only be usefully and properly interpreted if the speed at which it propagates never exceeds the speed of light.

In this chapter we explore a recent quantisation of the one-dimensional free EM field in the position representation [45]. The main focus of this quantisation will be the construction of single-photon wave packets in a basis of localised photonic excitations. We determine an equation of motion for these excitations which leads to a Schrödinger equation for the photon. By focussing on dynamics, new parameters are introduced that were previously neglected or overlooked. This provides us with a fuller description of the quantised EM field. For completeness, expressions for the EM field observables shall be constructed, and a comparison with standard quantisations shall be given.

2. The classical EM field in one dimension

The equations of motion for light are given by Maxwell's equations. The solutions to these equations provide us with the expected dynamics of the quantised particles of the EM field. The purpose of this section is to review the appropriate equations of motion and their solutions in one dimension, and to determine an expression for the energy of the EM field.

2.1 The dynamics of the EM field

2.1.1 Maxwell's equations

Light consists of two real, mutually propagating vector fields: the electric field and the magnetic field. In one dimension, the electric and magnetic fields propagate along a single axis parametrised by a position coordinate x. The electric and magnetic fields at a position x at a time t are denoted $\mathbf{E}(x, t)$ and $\mathbf{B}(x, t)$ respectively.

Although $\mathbf{E}(x, t)$ and $\mathbf{B}(x, t)$ are parametrised by a position along the x-axis only, the fields are oriented, or polarised, in the plane orthogonal to the direction of propagation. By specifying a right-handed Cartesian coordinate system (x, y, z), the electric and magnetic field have components in the y and z directions only. The components of the fields oriented along the y (z)-axis shall be referred to as horizontally (vertically) polarised. The polarisation of the field is specified by a discrete parameter $\lambda = \mathrm{H}, \mathrm{V}$.

In a dielectric medium of constant permittivity ε and permeability μ, and by denoting $c = (\varepsilon\mu)^{-1/2}$, the horizontally and vertically polarised components of $\mathbf{E}(x, t)$ and $\mathbf{B}(x, t)$ satisfy the following simplified forms of Maxwell's equations:

$$\begin{aligned} \frac{\partial}{\partial x}\mathrm{E}(x,t) &= \pm\frac{\partial}{\partial t}\mathrm{B}(x,t) \\ c^2\frac{\partial}{\partial x}\mathrm{B}(x,t) &= \pm\frac{\partial}{\partial t}\mathrm{E}(x,t) \end{aligned} \tag{1}$$

In both lines of Eq. (1) above, the electric and magnetic fields have alternate polarisations. The positive (negative) sign applies when the electric and magnetic fields are vertically (horizontally) and horizontally (vertically) polarised respectively.

2.1.2 The wave equation

Maxwell's equations, Eq. (1) couple together different components of the electric and magnetic field vectors. By combining these equations, we can construct a second-order differential equation for each of the four field components independently. These equations are

$$\begin{aligned} \left[\frac{\partial^2}{\partial x} - \frac{1}{c^2}\frac{\partial^2}{\partial t}\right]\mathbf{E}(x,t) &= 0 \\ \left[\frac{\partial^2}{\partial x} - \frac{1}{c^2}\frac{\partial^2}{\partial t}\right]\mathbf{B}(x,t) &= 0. \end{aligned} \tag{2}$$

Here we have four identical equations of motion, one for each of the four components of the EM field. The solutions to Eq. (2) will be examined in Section 2.3.

2.2 The energy and momentum of the EM field

2.2.1 The energy observable

At each point in space and time, the electric and magnetic fields exert a force on any charged matter present at that point. For this reason, the EM field is able to do mechanical work on charged matter, and must therefore store a certain amount of energy. Taking this into account, explicit expressions for the energy and momentum of light in a dielectric medium can be determined. By considering the work done by the fields on a charge current density in a dielectric medium, one can show that the total energy along the x-axis is given by the expression

$$H_{\text{energy}}(t) = \int_{-\infty}^{\infty} \mathrm{d}x \frac{A}{2} \left\{ \varepsilon |\mathbf{E}(x,t)|^2 + \frac{1}{\mu} |\mathbf{B}(x,t)|^2 \right\}. \tag{3}$$

Here A is the area occupied by the field in the y-z plane.

2.2.2 The Poynting vector

The energy stored in the EM field in a particular region is carried in the direction of propagation in the form of the Poynting vector **S**. Since in one dimension light can only propagate along the x-axis, the only non-zero component of the Poynting vector is the x component, which is given by the expression

$$S(x,t) = \frac{1}{\mu} [\mathrm{E_H}(x,t)\mathrm{B_V}(x,t) - \mathrm{E_V}(x,t)\mathrm{B_H}(x,t)]. \tag{4}$$

In the above the H and V subscripts refer to the horizontally and vertically polarised components of the fields respectively. The expressions above, particularly Eq. (3), will be of importance in Section 4.2.2.

2.3 The solutions to Maxwell's equations

2.3.1 Left- and right-propagating waves

The wave equation, Eq. (2) describes the propagation of a wave along the x-axis at a constant speed c. This is the speed of light in the medium. In only one dimension, the solutions of the wave equation take a simple form. By considering first the components of the electric field $\mathrm{E}_\lambda(x,t)$, one can show that the expressions

$$\mathrm{E}_\lambda(x,t) = \sum_{s=\pm 1} \mathrm{E}_{s\lambda}(x,t) \tag{5}$$

satisfy Eq. (2) when $\mathrm{E}_{s\lambda}(x,t) = \mathrm{E}_{s\lambda}(x - sct, 0)$. In Eq. (5) above, the parameter $s = \pm 1$ is introduced in order to differentiate between solutions propagating to the left (decreasing x) or the right (increasing x). In this notation, light characterised by $s = -1(+1)$ propagates to the left (right). The exact form of $\mathrm{E}_{s\lambda}(x,t)$ is determined from the initial conditions of the system.

2.3.2 Complete electric and magnetic field solutions

The corresponding magnetic field solution to Eq. (2) is not independent of the electric field solution. By using Maxwell's equation, Eq. (1), the magnetic field can be determined directly from the electric field solution (5). After taking into account the sign difference for different polarisations, one may show that

$$\mathbf{E}(x,t) = \sum_{s=\pm 1} c\left[\mathrm{E}_{s\mathrm{H}}(x,t)\hat{\boldsymbol{y}} + \mathrm{E}_{s\mathrm{V}}(x,t)\hat{\boldsymbol{z}}\right] \tag{6}$$

and

$$\mathbf{B}(x,t) = \sum_{s=\pm 1} s\left[-\mathrm{E}_{s\mathrm{V}}(x,t)\hat{\boldsymbol{y}} + \mathrm{E}_{s\mathrm{H}}(x,t)\hat{\boldsymbol{z}}\right]. \tag{7}$$

Here $\hat{\boldsymbol{y}}$ and $\hat{\boldsymbol{z}}$ are unit vectors oriented in the positive y and z directions respectively.

2.3.3 Energy and momentum

Since $\mathbf{E}(x,t)$ and $\mathbf{B}(x,t)$ are both characterised by the solutions $\mathrm{E}_{s\lambda}(x,t)$, the energy and Poynting vector of the field must also be characterised by these solutions. Substituting Eqs. (6) and (7) into Eqs. (3) and (4) one finds that

$$H_{\text{energy}}(t) = \sum_{s=\pm 1}\sum_{\lambda=\mathrm{H,V}} \int_{-\infty}^{\infty} \mathrm{d}x \ \frac{A}{2}\left\{\varepsilon\left|\mathrm{E}_{s\lambda}(x,t)\right|^2\right\} \tag{8}$$

and

$$S(x,t) = \sum_{s=\pm 1}\sum_{\lambda=\mathrm{H,V}} (sc/\mu)\left|\mathrm{E}_{s\lambda}(x,t)\right|^2. \tag{9}$$

It is clear from Eq. (9) that a positive Poynting vector indicates propagation to the right whereas a negative Poynting vector indicates propagation to the left.

3. The wave function of the photon

The equations of motion for light in a homogeneous and isotropic dielectric medium apply the first set of constraints to the particle behaviour of light. For a correct and natural interpretation of the photon wave function, the probability distribution of particles represented by the wave function must evolve identically to the classical wave packets of an EM wave. In this section we construct a Fock space of localised bosonic excitations that provide a basis for constructing single-photon wave packets. By imposing a constraint on the dynamics of these excitations in 1 + 1-dimensional space-time, a Schrödinger equation is formulated for the photon.

3.1 The parameter space of single-photon wave packets

3.1.1 Unitary time evolution

Consider the propagation of a photon wave packet through the dielectric medium along the x-axis. At an initial time $t = 0$, we may represent this wave packet in the Hilbert space by a state vector $|\psi_1(0)\rangle$. After a time t has passed, the photon wave packet is now found in the time-evolved state

$$|\psi_1(t)\rangle = U(t,0)|\psi_1(0)\rangle \tag{10}$$

where $U(t, 0)$ is the unitary time-evolution operator from time $t = 0$ to time t. As we have determined that light must propagate along the x-axis at a speed c, the unitary operator $U(t, 0)$ transports the left- and right-moving components of the wave packet to the left or the right by an exact distance ct.

3.1.2 A complete parameter space

If we consider two single-photon wave packets that are entirely distinguishable from each other, then their corresponding state vectors must be orthogonal. When we localise two photons to different points along the x-axis, they are distinguishable from each other. Localised photon states, therefore, are orthogonal to one another, and it is natural to characterise them by their position along the x-axis. In the same way, photons with different polarisations are distinguishable and their state vectors orthogonal. Photon states are therefore also characterised by a polarisation λ. In addition to this, states describing propagation in opposite directions must be orthogonal to one another, and must be characterised by the discrete parameter $s = \pm 1$.

To see that this last parametrisation must be so, consider the setup illustrated in **Figure 1** showing two identical single-photon wave packets propagating in opposite directions. We denote the state vectors for the left- and right-hand systems $|\psi_1(x,t)\rangle$ and $|\psi_2(x,t)\rangle$ respectively where $|\psi_1(x,0)\rangle = |\psi_2(x+2a,0)\rangle$. At an initial time $t = 0$, the photon in the left-hand diagram is localised to a position $x = -a$ whereas the photon in the right-hand diagram is localised to the position $x = a$. Since the two wave packets occupy separate regions of the x-axis, their state vectors must be orthogonal:

$$\langle\psi_1(x,0)|\psi_2(x,0)\rangle = \langle\psi_1(x,0)|\psi_1(x-2a,0)\rangle = 0. \tag{11}$$

At a later time $t = a/c$, both photons will have travelled a distance a to the left or the right of their initial positions. After taking into account the direction of propagation of each of the photons, at this later time $t = a/c$, both wave packets will coincide with each other perfectly at the origin. When parametrised by only a position and

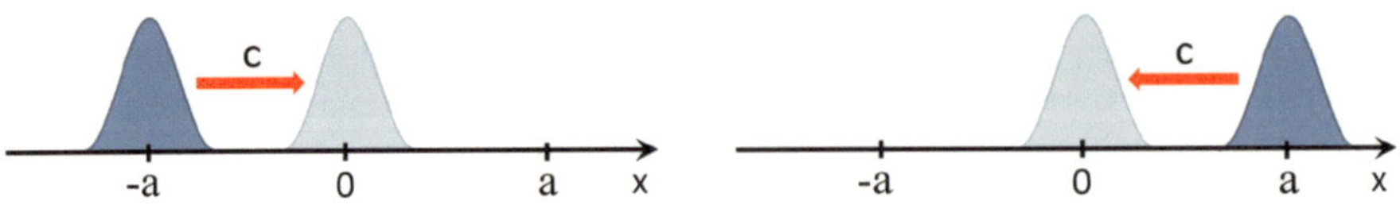

Figure 1.
The diagram illustrates the propagation of two localised wave packets. In the left-hand diagram, a single photon propagates to the right from an initial position $x = -a$. In the right-hand diagram, a single photon of identical shape propagates to the left from an initial position $x = a$. At a later time both wave packets reach the origin. Here the two wave packets are completely identical with respect to their position.

polarisation, therefore, their corresponding state vectors will no longer be orthogonal. Hence, the inner product

$$\langle \psi_1(x,t) | \psi_2(x,t) \rangle = \langle \psi_1(x,t) | \psi_1(x-2a,t) \rangle \tag{12}$$

will necessarily be non-zero. Given that the states $|\psi_1(x,t)\rangle$ and $|\psi_2(x,t)\rangle$ evolve unitarily according to Eq. (10), however, the inner product between the two states at a time t is given by

$$\begin{aligned} \langle \psi_1(x,t) | \psi_2(x,t) \rangle &= \langle \psi_1(x,0) | U^\dagger(t,0) U(t,0) | \psi_2(x,0) \rangle \\ &= \langle \psi_1(x,0) | \psi_2(x,0) \rangle \end{aligned} \tag{13}$$

where $\dagger$ denotes hermitian conjugation. This inner product must therefore be constant with respect to time. The assumption that our two state vectors are initially orthogonal, therefore, is inconsistent with unitary time evolution and we reach a contradiction. The resolution to this problem is to ensure that wave packets propagating in different directions remain orthogonal at all times. To properly differentiate between states propagating in different directions, therefore, single-photon wave packets must also be parametrised by $s = \pm 1$ in addition to position and polarisation.

3.2 Local photons

3.2.1 Creation and annihilation operators

During the interactions between light and matter, atoms absorb and emit light on the level of single photons. The appropriate Hilbert space for the free EM field, therefore, is a Fock space of identical and non-interacting bosonic particles. From what we have determined in the previous section, the localised photonic excitations of the EM field are characterised at any one time by a coordinate $x \in (-\infty, \infty)$, a polarisation $\lambda = \mathrm{H}, \mathrm{V}$ and a direction of propagation $s = \pm 1$. From now on we shall refer to such excitations as blips, which is the acronym for bosons localised in position.

As is usual for a system of identical particles we may define a collection of blip annihilation operators that remove a single blip from the system. The blip annihilation operator is denoted $a_{s\lambda}(x,t)$ in the Heisenberg picture and $a_{s\lambda}(x,0)$ in the Schrödinger picture. A state containing only a single blip is defined

$$|1_{s\lambda}(x,t)\rangle = a_{s\lambda}^\dagger(x,t)\, |0\rangle. \tag{14}$$

The operator $a_{s\lambda}^\dagger(x,t)$ is termed the blip creation operator and generates a single blip characterised by the parameters (x,t,λ,s). In Eq. (14) above, $|0\rangle$ is the vacuum state containing precisely zero blips. The vacuum state satisfies the property

$$a_{s\lambda}(x,t)\, |0\rangle = 0 \tag{15}$$

for all x, t, s and λ.

3.2.2 Commutation relations

Although the state defined in Eq. (14) contains only one blip, states containing an arbitrary number of blips can be generated by repeatedly applying the blip creation operators to the vacuum state. Since blips are bosons, the resulting state must be unchanged through any reordering of the blips' positions. Consequently, the ordering of these creation operators must be insignificant and they must commute with one another. Hence

$$\left[a_{s\lambda}^{\dagger}(x,t), a_{s'\lambda'}^{\dagger}(x',t')\right] = 0 = \left[a_{s\lambda}(x,t), a_{s'\lambda'}(x',t')\right] \tag{16}$$

for any $x, x', t, t', \lambda, \lambda'$, s and s'.

In Section 3.1.2 it was discussed how blips located at different positions, carrying different polarisations or propagating in opposite directions must be perfectly distinguishable from one another. As a consequence, the states that represent them must also be orthogonal. Taking this into account, we specify the following inner product for two single-blip states:

$$\langle 1_{s\lambda}(x,t) | 1_{s'\lambda'}(x',t)\rangle = \delta_{s,s'}\,\delta_{\lambda,\lambda'}\,\delta(x-x'). \tag{17}$$

Using Eqs. (14) and (15), and expressing the inner product (17) in terms of blip creation and annihilation operators, it can be shown that at any fixed time t

$$\left[a_{s\lambda}(x,t), a_{s'\lambda'}^{\dagger}(x',t)\right] = \delta_{s,s'}\,\delta_{\lambda,\lambda'}\,\delta(x-x'). \tag{18}$$

This is the fundamental commutation relation for blips.

3.2.3 The photon wave function

In the context of linear optics experiments [46, 47], it is usual to talk about single photons when referring to particles whose state vectors $|1(t)\rangle$ can be expressed $|1(t)\rangle = a^{\dagger}(t)\,|0\rangle$ where $a(t)$ is an annihilation operator satisfying the commutation relation

$$\left[a(t), a(t)^{\dagger}\right] = 1. \tag{19}$$

The blip states defined in Eq. (14) are not normalisable, but when superposed over a region of the x-axis can provide a localised basis for normalised single-photon wave packets. Taking this into account, the annihilation operator for a single-photon wave packet can be defined in the following way:

$$a(t) = \sum_{s=\pm 1}\sum_{\lambda=\mathrm{H,V}}\int_{-\infty}^{\infty} \mathrm{d}x\, \psi_{s\lambda}^{*}(x)\, a_{s\lambda}(x,t) \tag{20}$$

where $*$ denotes complex conjugation. In Eq. (20), the operator $a(t)$ is properly normalised and satisfies Eq. (19) when

$$\sum_{s=\pm 1}\sum_{\lambda=\mathrm{H,V}}\int_{-\infty}^{\infty} \mathrm{d}x\;\; |\psi_{s\lambda}(x)|^2 = 1. \tag{21}$$

The function $\psi_{s\lambda}(x,t)$ in Eq. (20) represents the probability amplitude for finding a photon with polarisation λ propagating in the s direction at a position x. More specifically, the transition probability between the single-photon state $|1(t)\rangle$ and the state $|1_{s\lambda}(x,t)\rangle$ is given by the expression

$$\left|\langle 0|a_{s\lambda}(x,t)a^{\dagger}(t)|0\rangle\right|^2 = |\psi_{s\lambda}(x)|^2. \tag{22}$$

Hence $\psi_{s\lambda}(x,t)$ has the correct properties to be correctly interpreted as a single-photon wave function in the position representation.

3.3 A Schrödinger equation for light

3.3.1 A Hamiltonian constraint

In order to calculate the dynamics of a quantum system, it is usual to first determine the Hamiltonian for that system. In a closed system, the Hamiltonian would be given by the energy observable. Once found, the Hamiltonian is used to construct a Schrödinger equation for state vectors in the Hilbert space. So far, an energy observable has not been constructed for the blip states. Moreover, there is no immediate choice for this observable, as, having complete uncertainty in their frequency, blips are not the eigenstates of the energy observable. Fortunately, however, the dynamics of single blips have already been determined. They are given by the solutions to Maxwell's equations, Eq. (5).

Blips are characterised by both a coordinate x in space and a coordinate t in time. A single blip, therefore, may exist at one position at one moment in time, and then at a different position at another moment in time. The classical dynamics of light in the medium places a constraint on which positions the blip may take from one moment to the next. Being more specific, in order to satisfy Maxwell's equations, the expectation value of a localised blip must propagate at a speed c along the x-axis without any dispersion. These dynamics are imposed by the constraint $\langle a_{s\lambda}(x,t)\rangle = \langle a_{s\lambda}(x-sct,0)\rangle$. Since this applies for any time-independent state, we can determine the general constraint

$$a_{s\lambda}(x,t) = a_{s\lambda}(x-sct,0). \tag{23}$$

When allowed to propagate freely, a blip found at x at a time t will be found at a position $x - sct$ at the time $t = 0$.

The constraint on the dynamics, Eq. (23), enables us to define an equation of motion for the blip operators $a_{s\lambda}(x,t)$. More specifically, by taking the time derivative of Eq. (23) it can be shown that

$$\left[\frac{\partial}{\partial t} + sc\frac{\partial}{\partial x}\right]a_{s\lambda}(x,t) = 0. \tag{24}$$

The equation above takes the form of a Wheeler-deWitt equation, and defines a stationary or "timeless" state of the system [48]. In the system considered here, this equation confines the trajectories of blips to the boundaries of the light cone. By relating a change in time to a change in the position of a blip in this way, we obtain a Schrödinger equation for blips:

$$i\hbar\frac{\partial}{\partial t}|1_{s\lambda}(x,t)\rangle = -i\hbar sc\frac{\partial}{\partial x}|1_{s\lambda}(x,t)\rangle. \tag{25}$$

3.3.2 The dynamical Hamiltonian

In the context of a Schrödinger equation, the motion of the blip given by the right-hand side of Eq. (25) is generated by the Hamiltonian for this system. It is very convenient to determine this Hamiltonian as it provides a basis for introducing interactions in more complex models. To this end, by using the Schrödinger equation for blips (25), we can determine exactly the Hamiltonian for the free propagation of light in a one-dimensional dielectric medium. Here we shall denote this operator $H_{\text{dyn}}(t)$ with the subscript "dynamical" to distinguish it as the Hamiltonian operator present in the Schrödinger equation.

Looking again at the right-hand side of Eq. (25), it can be seen that the number of blips, their polarisation and their direction of propagation are all preserved as they evolve in time. It is only their position that changes. Taking this into account, a suitable ansatz for the dynamical Hamiltonian $H_{\text{dyn}}(t)$ would be

$$H_{\text{dyn}}(t) = \sum_{s=\pm 1}\sum_{\lambda=\text{H},\text{V}} \int_{-\infty}^{\infty} \text{d}x \int_{-\infty}^{\infty} \text{d}x \;\; i\hbar sc f_{s\lambda}(x, x') a_{s\lambda}^{\dagger}(x, t) a_{s\lambda}(x', t) \tag{26}$$

where $f_{s\lambda}(x, x')$ is a function to be determined. This operator takes the form of an exchange operator that annihilates a blip at one position and replaces it with an identical blip at a different position. To ensure that $H_{\text{dyn}}(t)$ is hermitian, $f_{s\lambda}(x, x') = -f_{s\lambda}(x', x)$.

In the Heisenberg picture, the dynamics of a blip operator $a_{s\lambda}(x, t)$ can be equivalently expressed through Heisenberg's equation of motion:

$$\frac{\partial}{\partial t} a_{s\lambda}(x, t) = -\frac{i}{\hbar} \left[a_{s\lambda}(x, t), H_{\text{dyn}}(t) \right]. \tag{27}$$

Hence by substituting the Hamiltonian (26) into Heisenberg's Eq. (27), making use of the commutation relations (16) and (18), and ensuring equivalence to Eq. (24), it can be shown that

$$f_{s\lambda}(x, x) = -\frac{\partial}{\partial x} \delta(x - x) \tag{28}$$

and therefore

$$H_{\text{dyn}}(t) = -i \sum_{s=\pm 1}\sum_{\lambda=\text{H},\text{V}} \int_{-\infty}^{\infty} \text{d}x \;\; \hbar sc\, a_{s\lambda}^{\dagger}(x, t) \frac{\partial}{\partial x} a_{s\lambda}(x, t). \tag{29}$$

This Hamiltonian is hermitian and therefore a generator of unitary dynamics. It should also be noted that the Hamiltonian for right-propagating blips takes the negative value of the Hamiltonian for left-propagating blips. This demonstrates that a right-propagating blip behaves identically to a left-propagating blip when the direction of time is reversed.

4. Field observables in the position representation

The approach to quantisation taken here differs from usual procedures by focussing on the particle character of the EM field rather than the quantised field observables. It is by taking this point of view that the field observables do not require a direct relationship to the wave function of the photon. This view is also held in Ref. [49]. The field observables remain, however, the fundamental observables from which we may derive expressions for the energy and momentum of the EM field. The purpose of this section is to construct the electric and magnetic field observables in the position representation acting on the extended blip Hilbert space. By insisting that blips are the localised excitations of the EM field, the field observables obtain unique characteristics that are crucial for a fuller understanding of many quantum effects.

4.1 The EM field observables

4.1.1 An ansatz for the EM field observables

The electric and magnetic field observables $\mathbf{E}(x, t)$ and $\mathbf{B}(x, t)$ are a linear and hermitian superposition of the creation and annihilation operators for the photonic excitations of the system. In the position representation, these are the blip operators $a^{\dagger}_{s\lambda}(x,t)$ and $a_{s\lambda}(x,t)$. Although this superposition is linear, there is no reason to assume that this superposition must be local. In other words, the field observables at a position x do not need to be a superposition of blip operators defined at that same point only. For this reason, it is useful to introduce the notation

$$\mathcal{R}[a_{s\lambda}](x,t) = \int_{-\infty}^{\infty} \mathrm{d}x' \ \mathcal{R}_{s\lambda}(x,x')a_{s\lambda}(x',t). \tag{30}$$

Here $\mathcal{R}$ is referred to as the regularisation operator and $\mathcal{R}(x,x')$ is a distribution over the x axis.

In the following, the operators $\mathbf{E}(x, t)$ and $\mathbf{B}(x, t)$ shall denote the complex part of the electric and magnetic field observables respectively. The total real fields are given by the hermitian superposition $(\mathbf{O} + \mathbf{O}^{\dagger})/2$ where $\mathbf{O} = \mathbf{E}, \mathbf{B}$. Taking this into account, an appropriate ansatz for the complex field observables is

$$\mathbf{E}(x,t) = \sum_{s=\pm 1} c\left[\mathcal{R}[a_{s\mathrm{H}}](x,t)\hat{\boldsymbol{y}} + \mathcal{R}[a_{s\mathrm{V}}](x,t)\hat{\boldsymbol{z}}\right] \tag{31}$$

and

$$\mathbf{B}(x,t) = \sum_{s=\pm 1} s\left[-\mathcal{R}[a_{s\mathrm{V}}](x,t)\hat{\boldsymbol{y}} + \mathcal{R}[a_{s\mathrm{H}}](x,t)\hat{\boldsymbol{z}}\right]. \tag{32}$$

It may be noted here that all components of the real electric and magnetic field observables commute.

4.1.2 The regularisation operator

The regularisation operator $\mathcal{R}$ provides a relationship between the field observables and the blip operators. Whilst in many quantisations photon wave packets must be locally related to the field observables, for a general choice of $\mathcal{R}(x,x')$, blips at one position may contribute to the field observables at another position. In fact, we shall see later in this chapter that a single blip contributes to the field observables at all positions along the x-axis. Notwithstanding this, the function $\mathcal{R}(x,x')$ must satisfy several general conditions.

Like the blip operators, the expectation values of the field observables must satisfy Maxwell's equations with respect to any time-independent state. Taking into account the orientation of the field components in Eqs. (31) and (32), this condition implies that the regularised blip operators $\mathcal{R}[a_{s\lambda}](x,t)$ must satisfy Eq. (24). Since this equation is also satisfied by the blip operators, using Eq. (30) it can be demonstrated that $\mathcal{R}_{s\lambda}(x,x')$ must be position invariant; that is, $\mathcal{R}_{s\lambda}(x,x') = \mathcal{R}_{s\lambda}(x-x')$. What is more, since the medium is homogeneous and isotropic, the regularisation must be symmetric, $\mathcal{R}_{s\lambda}(x-x') = \mathcal{R}_{s\lambda}(x'-x)$, and independent of s and λ, $\mathcal{R}_{s\lambda}(x-x') = \mathcal{R}(x-x')$.

4.2 Energy in the position representation

4.2.1 The energy observable

Now that we have a pair of expressions for the electric and magnetic field observables, it is possible to determine the energy observable for the free field in one-dimension. To do so we substitute the field observables (31) and (32) into the classical expression for the energy determined in Eq. (3). In return we find that

$$H_{\text{energy}}(t) = \sum_{s=\pm 1}\sum_{\lambda=\text{H},\text{V}} \int_{-\infty}^{\infty} \mathrm{d}x \ \frac{A\varepsilon c^2}{4} \{\mathcal{R}[a_{s\lambda}](x,t) + H.c\}^2 \tag{33}$$

Due to the square in the integrand, this observable is strictly positive as would be expected for an energy. This result, however, implies that the energy observable cannot be equal to the dynamical Hamiltonian. Whereas the energy of a single blip is always positive, the left- and right-moving components of the dynamical Hamiltonian have opposite signs.

4.2.2 Energy conservation

In a closed system, energy is always conserved. In standard quantisations, when the dynamical Hamiltonian is equivalent to the energy observable, conservation of energy is guaranteed automatically as a consequence of Heisenberg's equation. We have seen, however, that in this quantisation the dynamical Hamiltonian and the energy observable are not equal. Energy conservation is only guaranteed, therefore, if $H_{\text{energy}}(t)$ and $H_{\text{dyn}}(t)$ commute. Using the expressions for $H_{\text{dyn}}(t)$ and $H_{\text{energy}}(t)$ given in Eqs. (26) and (33) respectively, and by taking into account that $f_{s\lambda}(x-x')$ is an odd function and $\mathcal{R}(x-x')$ an even function, it can be shown that the dynamical Hamiltonian and the energy observable commute with each other. Hence, the energy of the free EM field is conserved.

4.3 Non-local contributions to the field observables

4.3.1 Monochromatic excitations

It can be seen from Eq. (33) that the regularisation operator plays an important role in determining the energy of a photon. Since the energy of a photon is determined by its frequency, it is convenient to express the energy observable (33) in a basis of monochromatic excitations. Such a set of excitations can be constructed by considering the Fourier transform of the localised blip operators. We introduce, therefore, the operators

$$a_{s\lambda}(k,t) = \int_{-\infty}^{\infty} \frac{\mathrm{d}k}{\sqrt{2\pi}} \; e^{-ikx} a_{s\lambda}(x,t) \tag{34}$$

which, using Eq. (18), can be shown to satisfy the equal-time commutation relation

$$\left[a_{s\lambda}(k,t), a^{\dagger}_{s'\lambda'}(k',t)\right] = \delta_{ss'}\,\delta_{\lambda\lambda'}\,\delta(k-k'). \tag{35}$$

All annihilation operators commute amongst themselves, as do the creation operators.

4.3.2 The energy of a photon

Expressed in terms of the monochromatic operators $a_{s\lambda}(k,t)$ and their hermitian conjugates, the energy observable (33) takes the alternative form

$$H_{\text{energy}}(t) = \sum_{s=\pm 1} \sum_{\lambda=\text{H},\text{V}} \int_{-\infty}^{\infty} \mathrm{d}k \; \frac{A\varepsilon\pi c^2}{2} \|\mathcal{R}(k)\, a_{s\lambda}(k,t) + \mathcal{R}^*(-k)\, a^{\dagger}_{s\lambda}(-k,t)\|^2. \tag{36}$$

In the expression above, $\mathcal{R}(k)$ is the Fourier transform of the regularisation function $\mathcal{R}(x-x')$ defined earlier. By taking into account the commutation relation (35), we can show that the energy expectation value of a single monochromatic excitation of angular frequency $\omega = skc$ is

$$\langle 1_{s\lambda}(k,t)|H_{\text{energy}}(t)|1_{s\lambda}(k,t)\rangle = A\varepsilon\pi c^2\, |\mathcal{R}(k)|^2\, \delta(0). \tag{37}$$

In the above the state $|1_{s\lambda}(k,t)\rangle$ is defined analogously to the blip state in Eq. (14). The delta function appearing in Eq. (37) is due only to the infinite normalisation of the monochromatic states.

When an atom with transition energy $\hbar|\omega|$ emits a photon, exactly one excitation is generated oscillating with an angular frequency ω. If the total energy is carried away by the photon, an excitation of frequency ω must have an energy $\hbar|\omega|$. Equating the energy of the monochromatic excitation, therefore, with the expectation value (37), we can determine, up to an overall phase, an expression for the function $\mathcal{R}(k)$. Doing so we find that

$$\mathcal{R}(k) = \sqrt{\frac{\hbar|k|}{A\varepsilon\pi c}} \tag{38}$$

4.3.3 A non-local regularisation

Now that we have determined $\mathcal{R}(k)$, we can calculate $\mathcal{R}(x-x')$ explicitly, thus providing us with a relationship between blips and the field observables in the position representation. Taking the Fourier transform of Eq. (38), we find that

$$\mathcal{R}(x-x') = \int_{-\infty}^{\infty} \frac{\mathrm{d}k}{2\pi} \; e^{ik(x-x')} \sqrt{\frac{2\hbar|k|}{\varepsilon A c}}. \tag{39}$$

When $x \neq x'$, this expression can be given in the alternative form

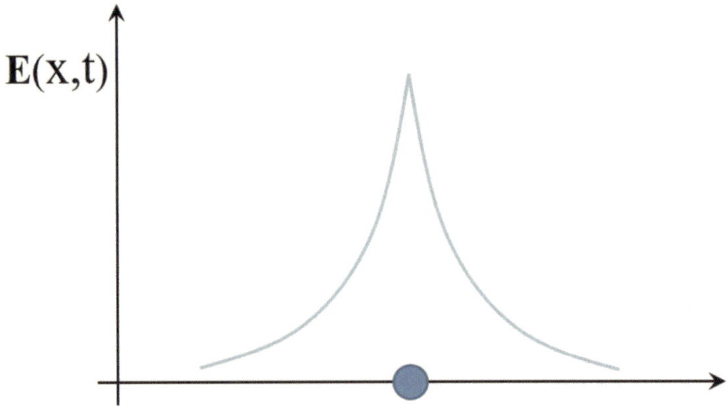

Figure 2.
The figure illustrates the contribution of a single blip (the blue spot) to the electric field observable. The blip in the diagram contributes to measurements of the electric field observable at all points along the x axis. The magnitude of the contribution decreases with the distance of the measurement from the blip.

$$\mathcal{R}(x - x') = -\sqrt{\frac{\hbar}{4\pi\varepsilon Ac}}\frac{1}{|x - x'|^{3/2}}. \qquad (40)$$

The distribution $\mathcal{R}(x - x')$ is non-vanishing everywhere, and decreases away from the origin with the negative three halves power of distance.

Whilst the blips represent the localised particles of the system, and interact only locally with their surroundings, they contribute to the measurement of the electric and magnetic field observables in a highly non-local way. One way to think about this relationship is that the measurement of the electric and magnetic fields takes into account the behaviour of blips at all positions in space. An alternative perspective would be to treat each blip as a "carrier" of a non-local field. This interpretation is visualised in **Figure 2** showing the contribution of a single blip to the electric field. In a similar manner to as one may think about the gravitational field about the earth, one may think of a single blip as being surrounded by and carrying with it a non-local electromagnetic field.

5. Conclusions

In standard treatments of the quantised EM field in one dimension, single photon wave packets are characterised by a wave vector k and a polarisation λ. By demanding that photons can be localised, however, and that they must propagate at the speed of light along the x-axis, we have shown that an additional discrete parameter $s = \pm 1$ must be introduced, thus doubling the size of the usual Hilbert space. Whilst canonical quantisation overlooks these degrees of freedom, by taking the viewpoint of a Schrödinger equation, these additional degrees of freedom arise naturally. What is more, by maintaining the perspective of a Schrödinger equation, a Hamiltonian operator $H_{\mathrm{dyn}}(t)$ was constructed that generates the time-evolution of state vectors in the Hilbert space. Unlike other quantisations, however, our Hamiltonian necessarily possess both positive and negative eigenvalues. This can be most clearly seen by expressing the dynamical Hamiltonian in the momentum representation. In terms of the monochromatic operators defined in Eq. (34), $H_{\text{dyn}}(t)$ is given by the expression

$$H_{\text{dyn}}(t) = \sum_{s=\pm 1} \sum_{\lambda=\text{H},\text{V}} \int_{-\infty}^{\infty} \text{d}k \;\; \hbar sck \, a^{\dagger}_{s\lambda}(k,t) a_{s\lambda}(k,t). \tag{41}$$

This Hamiltonian is almost identical to the usual Hamiltonian of the free EM field, but the eigenvalues $\hbar sck$ can take both positive and negative values.

In standard quantisations, the Hamiltonian is always positive. The Hamiltonian above is positive only when the signs of s and k coincide, or equivalently when the sign of k indicates direction of propagation. Under this condition the dynamical Hamiltonian (41) and the energy observable (36) are equivalent. In this quantisation we therefore go beyond current assumptions by demanding that the Hamiltonian can be both positive and negative. This idea is not new, however. In the work of Hegerfeldt and others, it was shown that a wave packet cannot propagate causally when the Hamiltonian is bounded from below [37]. Mostafazadeh and Zamani [50], therefore, introduced a new inner product that enabled the use of negative frequency states. More recently, with similar justifications, Hawton considers real EM field excitations that necessarily contain both positive and negative frequency contributions [51–53].

By extending the Hilbert space to include both positive- and negative- frequency photons, it is possible to localise a photon not only in space, but also in time. In quantum physics, it has been a significant problem to define a time operator [48]. Whereas the position of a particle is associated with a position observable, time is only a parameter of the system. Since in our quantisation all wave packets propagate causally, an operator can be defined that determines the time at which a particle will reach a particular position. In investigations into a "quantum relativity", where both space and time are placed on an equal quantum footing, a key concept is that of a closed and stationary universe [54]. The dynamics of these systems are constrained by a Wheeler-deWitt equation. As the dynamical constraint for blips (24) is of this form, and defines a stationary state in the global Hilbert space of excitations in space-time, the blip quantisation may provide a useful and insightful scheme for the modelling of quantum clocks and the study of a quantum time.

One of the most significant features of this quantisation is the non-locality between the blip operators and the field observables. This result is a direct consequence of the frequency-dependence of energy carried by the EM field. In many schemes, localisation of a photon is synonymous with localisation of the field observables. To fix the disparity between a probabilistic wave function and an energy carrying field, however, non-local or non-hermitian inner products are introduced. Although often elegant, such inner products can be cumbersome, and may depend upon the boundary conditions of the system. In the view of this quantisation, however, an excitation of the field is only localised if it satisfies the orthogonality relation (17), and such non-local contributions must therefore exist. Whilst this viewpoint may raise questions regarding instantaneous contributions to the field observables, the non-locality of the field observable is an important feature of the EM field. For instance, in Ref. [55], it is shown that these non-local field contributions are responsible for the Casimir effect between two parallel conducting plates.

The Schrödinger equation for light gives an alternative perspective on the particle behaviour of the free EM field. By returning to principles of wave-particle duality, the existing theory of the EM field has been shown to fall short with regard to the propagation of localised particles, and as a result new physics has been unearthed. The new quantisation that has followed provides a major change in our perspective on the interactions of photons and the way we describe them. For instance, using this quantisation one can construct a local interaction Hamiltonian for a double-sided semi-transparent mirror; something that was not previously possible [56]. Moreover, an investigation of blips in a cavity leads to a new perspective on the origin of the Casimir effect [55]. In future, this quantisation may contribute towards a fuller understanding of well-known quantum effects, and may also provide the tools necessary for studying new ones.

Acknowledgements

The author acknowledges financial support from the UK Engineering and Physical Sciences Research Council EPSRC [grant number EP/W524372/1]. The author also acknowledges the contributions of M. Basil Altaie, Almut Beige, Rob Purdy and Jake Southall to the original research that is presented and discussed within this chapter.

Conflict of interest

The author declares no conflict of interest.

Author details

Daniel R.E. Hodgson
University of Leeds, Leeds, United Kingdom

*Address all correspondence to: phydreh@leeds.ac.uk

References

[1] Bennett R, Barlow TM, Beige A. A physically-motivated quantisation of the electromagnetic field. European Journal of Physics. 2016;**37**(1):014001. DOI: 10.1088/0143-0807/37/1/014001

[2] Cohen-Tannoudji C, Dupont-Roc J, Grynberg G. Photons and Atoms: Introduction to Quantum Electrodynamics. Weinheim: WILEY-VCH Verlag GmbH Co. KGaA; 2004

[3] Newton TD, Wigner EP. Localized states for elementary systems. Reviews of Modern Physics. 1949;**21**(3):400-406

[4] Jordan TF. Simple proof of no position operator for quanta with zero mass and nonzero helicity. Journal of Mathematical Physics. 1978;**19**:1382

[5] Schwartz PK, Giulini D. Classical perspectives on Newton-Wigner position observable. International Journal of Geometric Methods in Modern Physics. 2020;**17**(12):2050176

[6] Hoffmann SE. No relativistic Newton-Wigner probability current for any spin. Journal of Physics A: Mathematical and Theoretical. 2019;**52**:225301

[7] Roswarne D, Sarkar S. Rigorous theory of photon localizability. Quantum Optics. 1992;**4**:405. DOI: 10.1088/0954-8998/4/6/005

[8] Jauch JM, Piron C. Generalized localizability. Helvetica Physica Acta. 1969;**42**:559-570

[9] Amrein WO. Localizability for particles of mass zero. Helvetica Physica Acta. 1969;**42**:149-190

[10] Pike ER, Sarkar S. Loop coherent states and photon localizability. Quantum Optics. 1993;**5**:251-256

[11] Chan KW, Law CK, Eberly JH. Localized single-photon wave functions in free space. Physical Review Letters. 2002;**88**:100402

[12] Wightman AS. On the localizability of quantum mechanical systems. Reviews of Modern Physics. 1962;**34**(4): 845-872

[13] Fleming GN. Covariant position operators, spin and locality. Physics Review. 1965;**139**(4B):188-197

[14] Fleming GN. Non-local properties of stable particles. Physics Review. 1965; **139**(4B):963-968

[15] Hawton M. Photon position operators with commuting components. Physical Review A. 1999;**59**(2):954-959

[16] Hawton M, Baylis WE. Photon position operators and localized bases. Physical Review A. 2001;**64**:012101

[17] Hawton M, Debierre V. Photon position eigenvectors, Wigner's little group and Berry's phase. Journal of Mathematical Physics. 2019;**60**:052104

[18] Landau L, Peierls R. Quantum electrodynamics in configuration space. Zeitschrift für Physik. 1930;**62**:188-200

[19] Cook RJ. Photon dynamics. Physical Review A. 1982;**25**:4

[20] Cook RJ. Local covariance of photon dynamics. Physical Review A. 1982;**26**(5): 2754-2760

[21] Białynicki-Birula I. Photon wave function. In: Wolf E, editor. Progress in Optics XXXVI. Amsterdam: Elsevier; 1996. pp. 245-294

[22] Białynicki-Birula I, Białynicki-Birula Z. The role of the Riemann-Silberstein vector in classical and quantum theories of electromagnetism. Journal of Physics A: Mathematical and Theoretical. 2013; **46**:053001

[23] Białynicki-Birula I. On the wave function of the photon. Acta Physica Polonica A. 1994;**86**(1–2):97-116

[24] Sipe JE. Photon wave functions. Physical Review A. 1995;**52**(3):1875-1883

[25] Smith BJ, Raymer MG. Photon wave functions, wave packet quantisation of light, and coherence theory. New Journal of Physics. 2007;**9**:414

[26] Knight JM. Strict localization in quantum field theory. Journal of Mathematical Physics. 1961;**2**:459

[27] Licht AL. Strict localization. Journal of Mathematical Physics. 1963;**4**:1443

[28] Licht AL. Local states. Journal of Mathematical Physics. 1966;**7**:1656

[29] Białynicki-Birula I, Białynicki-Birula Z. Why photons cannot be sharply localised. Physical Review A. 2009;**79**: 032112

[30] Southall J, Hodgson D, Purdy R, Beige A. Comparing Hermitian and non-Hermitian quantum electrodynamics. Symmetry. 2022;**14**(9):1816

[31] Hawton M. Photon position eigenvectors lead to complete photon wave mechanics. In: Proceedings of SPIE Vol. 6664, the Nature of Light: What Are Photons?; 26–30 August 2007. San Diego, California: SPIE; 2007. p. 666408

[32] Hawton M. Photon wave mechanics and position eigenvectors. Physical Review A. 2007;**75**:062107

[33] Gross L. Norm-invariance of mass zero equations under the conformal group. Journal of Mathematical Physics. 1964;**5**:687

[34] Hegerfeldt GC. Remark on causality and particle localization. Physical Review D. 1974;**10**(10):3320-3321

[35] Malament DB. In defense of dogma: Why there cannot be a relativistic quantum mechanics of localizable particles. In: Clifton R, editor. Perspectives on Quantum Reality: Non-relativistic, Relativistic and Field-Theoretic. 1st ed. Dordrecht: Springer; 1996. pp. 1-10

[36] Hegerfeldt GC. Instantaneous spreading and Einstein causality in quantum theory. Annals of Physics. 1998;**7**:716-725

[37] Hegerfeldt GC, Ruijsenaars SNM. Remarks on causality, localization, and spreading of wave packets. Physical Review D. 1980;**22**(2):377-384

[38] Halvorson H, Clifton R. No place for particles in relativistic quantum theories? Philosophy in Science. 2002; **69**(1):1-28

[39] Perez JF, Wilde IF. Localization and causality in relativistic quantum mechanics. Physical Review D. 1977; **16**(2):315-317

[40] Barat N, Kimball JC. Localization and causality for a free particle. Physics Letters A. 2003;**308**:110-115

[41] Biswas AK, Compagno G, Palma GM, Passante R, Porsico F. Virtual photons and causality in the dynamics of a pair of two-level atoms. Physical Review A. 1990;**42**(7):4291-4301

[42] Power EA, Thirunamachandran T. Analysis of causal behaviour in energy

transfer between atoms. Physical Review A. 1997;**56**(5):3395-3408

[43] Rubin MH. Violation of Einstein causality in a model quantum system. Physical Review D. 1987;**35**(12):3836-3839

[44] Valentini A. Non-local correlations in quantum electrodynamics. Physics Letters A. 1991;**153**(6–7):321-325

[45] Hodgson D, Southall J, Purdy R, Beige A. Local photons. Frontiers in Photonics. 2022;**3**:978855

[46] Żukowski M, Zeilinger A, Horne MA. Realizable higher-dimensional two-particle entanglement via multiport beam splitters. Physical Review A. 1997;**55**:2564

[47] Lim YL, Beige A. Generalized Hong-Ou-Mandel experiments with bosons and fermions. New Journal of Physics. 2005;**7**:155

[48] Altaie MB, Hodgson D, Beige A. Time and quantum clocks: A review of recent developments. Frontiers in Physics. 2022;**10**:897305

[49] Fleming GN. Reeh-Schlieder meets Newton-Wigner. Philosophy in Science. 2000;**67**:495

[50] Mostafazadeh A, Zamani F. Quantum mechanics of Klein-Gordon fields I: Hilbert space, localized states, and chiral symmetry. Annals of Physics. 2006;**321**:2183

[51] Hawton M. Validation of classical modeling of single-photon pulse propagation. Physical Review A. 2023; **107**:013711

[52] Hawton M. Maxwell quantum mechanics. Physical Review A. 2019;**100**: 012122

[53] Hawton M. Photon quantum mechanics in real Hilbert space. Physical Review A. 2021;**104**:052211

[54] Page DN, Wootters WK. Evolution without evolution: Dynamics described by stationary observables. Physical Review D. 1983;**27**:2885

[55] Hodgson D, Burgess C, Altaie MB, Beige A, Purdy R. An intuitive picture of the Casimir effect. arXiv:2203.14385 [quant-ph]. 2022:1-8

[56] Southall J, Hodgson D, Purdy R, Beige A. Locally acting mirror Hamiltonians. Journal of Modern Optics. 2021;**68**(12):647-660

Chapter 3

Path Integral of Schrödinger's Equation

Hocine Boukabcha, Salah Eddin Aid and Amina Ghobrini

Abstract

The path integral is a powerful tool for studying quantum mechanics because it has the merit of establishing the connection between classical mechanics and quantum mechanics. This formalism quickly gained prominence in various fields of theoretical physics, including its generalization to quantum field theory, quantum mechanics, and statistical physics. Using the Feynman propagator, we can calculate the partition function, the free energy, wave functions, and the energy spectrum of the considered physical system. Moreover, the Feynman formalism finds broad applications in geophysics and in the field of financial sciences.

Keywords: radial propagator, space–time transformation, modified Pöschl-Teller potential, energy spectrum, wave functions

1. Introduction

In this chapter, the Schrödinger solutions of potential problem have been evaluated using the Feynman path integral formulation of quantum mechanics; an appropriate space-time transformation has been applied to Green's function associated with the problem, which made it an integrable function. Also, the energy spectrum in a non-relativistic regime with normalized wave functions for potential, is obtained using path integral formalism of quantum mechanics; the results are evaluated for any state due to the use of an approximation scheme for centrifugal term $1/r^2$, the constructed propagator associated with the Schrödinger equation of the problem was treated by space-time transformation trick that made it integrable, and energy eigenvalues for some exceptional cases of potential were also presented to compare our solutions with those obtained in previous studies. The organization of this chapter is as follows: in Section 1, we formulate the radial propagator and its corresponding Green's function associated with a nonrelativistic particle in the presence of a potential where we use an approximation to the centrifugal term. In Section 2, we treat Green's function of Generalized inverse quadratic Yukawa potential by performing a nontrivial space-time transformation to pass from the actual complex problem to another already solved one, which is a Pöschl-Teller (PT) potential problem. In Section 3, energy eigenvalues and corresponding eigenfunctions are extracted from the poles and residues of the aforementioned solved Green's function. Section 4 discusses special cases of Deng Fun potentiel, Generalized inverse quadratic Yukawa potential as Kratzer potential, Yukawa potential, inversely quadratic Yukawa potential, and Coulomb potential.

2. Propagator and Schrödinger equation

The Schrödinger equation is a fundamental equation in quantum physics that describes the behavior of quantum systems. It was formulated by Erwin Schrödinger in 1925. The Schrödinger equation describes the time evolution of the wave function of a quantum system is governed by the equation:

$$i\hbar\frac{d}{dt}|\psi(t)\rangle = \hat{H}|\psi(t)\rangle, \tag{1}$$

which integrates in the particular case of a time-independent Hamiltonian:

$$|\psi(t')\rangle = \exp\left[-\frac{i}{\hbar}\hat{H}(t'-t)\right]|\psi(t)\rangle, \tag{2}$$

where $\hat{U}(t',t) = \exp\left[-\frac{i}{\hbar}\hat{H}(t'-t)\right]$ represents the evolution operator. Let us consider a particle in a potential, the Hamiltonian is written as:

$$\hat{H} = \frac{\hat{P}^2}{2m} + V(\hat{x}). \tag{3}$$

In the position representation x, the evolution equation becomes:

$$\langle x'|\psi(t')\rangle = \left\langle x'\right|\exp\left(\frac{-i}{\hbar}\hat{H}(t'-t)\right)\left|\psi(t)\right\rangle. \tag{4}$$

Let us use the position closure relation x

$$\int dx|x\rangle\langle x| = 1. \tag{5}$$

Eq. (4) becomes:

$$\langle x'|\psi(t')\rangle = \int dx\left\langle x'\right|\exp\left(\frac{-i}{\hbar}\hat{H}(t'-t)\right)\left|\psi(t)\right\rangle. \tag{6}$$

Thus, it can be written as

$$\psi(x',t') = \int K(x',t';x,t)\psi(x,t)dx. \tag{7}$$

The propagator $K(x',t';x,t) = \langle x'|\exp\left(\frac{-i}{\hbar}\hat{H}(t'-t)\right)|x\rangle$ allows us to evaluate the transition amplitude between the two states. Let us consider an initial state localized at x_0

$$\psi(x,t_0) = \delta(x-x_0), \tag{8}$$

then,

$$\psi(x',t') = K(x',t';x_0,t_0). \tag{9}$$

the probability amplitude of finding the particle at position x' at time t'.

3. Transition from propagator to Green's function

As we saw earlier, the propagator can be expressed in terms of the time-evolution operator as follows [1, 2]:

$$K(x'',t'';x',t') = \langle x''|U(t'',t')|x'\rangle, \tag{10}$$

where

$$U(t'',t') = \exp\left[\frac{-i}{\hbar}\hat{H}(t''-t')\right], \tag{11}$$

with $T = t'' - t'$. Moreover, it is possible to extract the energy spectrum as well as the wave function corresponding to a given physical system, from Green's function, the latter being none other than the Fourier transform of the propagator. In effect [3],

$$G(x'',x';E) = \frac{i}{\hbar}\int_0^{\infty} dT e^{i(E+i\varepsilon)T/\hbar} K(x'',x';T), \tag{12}$$

where ε, is a positive constant

$$(\hat{H} - E)G(x'',x';E) = \delta(x''-x'). \tag{13}$$

Formula (10) allows us to write:

$$G(x'',x';E) = \left\langle x''|\frac{1}{\hat{H} - E - i\varepsilon}|x'\right\rangle. \tag{14}$$

By introducing the closure relation on position $|n,\ell\rangle$, we can express the probability amplitude (11) as follows:

$$G(x'',x';E) = \frac{i}{\hbar}\int_0^{\infty}\sum_{n,\ell} dT\langle x''|n,\ell\rangle\left\langle n,\ell|e^{\frac{i(E+i\varepsilon)T}{\hbar}}e^{-\frac{iHT}{\hbar}}|x'\right\rangle, \tag{15}$$

or alternatively,

$$G(x'',x';E) = \sum_{n,\ell}^{\infty}\frac{\chi_{n,\ell}^{*}(x')\chi_{n,\ell}(x'')}{E_{n,\ell} - E - i\varepsilon}. \tag{16}$$

where $\chi_{n,\ell}(x)$ is the wave function corresponding to the eigenenergy $E_{n,\ell}$, and it is also possible to arrive at $\chi_{n,l}(x)$ and $E_{n,\ell}$ from the relation.

$$K(x'',t'';x',t') = \sum_{n,\ell}^{\infty} \chi^*_{n,\ell}(x')\chi_{n,\ell}(x'')e^{-\frac{iE_{n,\ell}T}{\hbar}}. \tag{17}$$

4. Path integral in spherical coordinates

In quantum mechanics, rotational symmetry is crucial in finding the wave functions and corresponding energies of physical systems. Spherical coordinates transform from the Schrödinger equation of rotational symmetry. Therefore, we can separate this equation into an angular part expressed in terms of spherical harmonics, whose solutions are known, and a radial part that contains specific information about the dynamical systems.

In the path integral, this coordinate transformation is possible, but initially, things become complicated. One of these complexities arises when studying the presence of a centrifugal barrier, which eliminates the possibility of "time slicing."

The following relation represents the formula for the three-dimensional (3D) propagator [4, 5]:

$$\begin{aligned} K\left(\vec{r''},t'';\vec{r'},t'\right) &= \int_{\vec{r'}}^{\vec{r''}} D\vec{r}(t)\exp\int\frac{i}{\hbar}\left(\frac{m}{2}(\Delta\vec{r_j})^2 - V\left(\vec{r}\right)\right)dt \\ &= \lim_{N\to\infty}\prod_{j=1}^{N+1}\left(\frac{m}{2i\pi\hbar\varepsilon}\right)^{\frac{1}{2}}\left[\prod_{i=1}^{N}\int_{R^3} d\vec{r}_j\right]\exp\left[\frac{i}{\hbar}S_N\right], \end{aligned} \tag{18}$$

with

$$t'' = t_{N+1};\ t' = t_0$$
$$\vec{r''} = \vec{r}_{N+1};\ \vec{r'} = \vec{r}_0$$

and the total action:

$$S_N = \sum_{j=1}^{N+1} S_j = \sum_{j=1}^{N+1}\left(\frac{m}{2\varepsilon}\left(r_j^2 + r_{j-1}^2 - 2\vec{r}_j\cdot\vec{r}_{j-1}\right) - \varepsilon V\left(\vec{r_j}\right)\right). \tag{19}$$

Using the spherical coordinate system (r,θ,φ) defined as:

$$\begin{cases} x = r\sin\theta\cos\varphi \\ y = r\sin\theta\sin\varphi, \\ z = r\cos\theta \end{cases} \tag{20}$$

with $r > 0; 0 \le \theta < \pi$ and $0 \le \varphi < 2\pi$.

Where the volume element is expressed in spherical coordinates as:

$$d\vec{r_j} = r_j^2\sin\theta_j dr_j d\theta_j d\varphi_j, \tag{21}$$

the propagator (18) can be rewritten in spherical coordinates as:

$$K\left(\vec{r''},t'';\vec{r'},t'\right) = \lim_{N\to\infty}\prod_{j=1}^{N+1}\left(\frac{m}{2i\pi\hbar\varepsilon}\right)^{\frac{3}{2}}\prod_{j=1}^{N}\left[\int_0^{\infty}\int_0^{\pi}\int_0^{2\pi} r_j^2\sin\theta_j dr_j d\theta_j d\varphi_j\right] \times \prod_{j=1}^{N+1}\left[\exp\left(\frac{i}{\hbar}S_j\right)\right], \tag{22}$$

The elemental action is:

$$S_j = \frac{m}{2\varepsilon}\left(r_j^2 + r_{j-1}^2 - 2r_jr_{j-1}\cos\Theta_{j,j-1}\right) - \varepsilon V\left(r_j\right), \tag{23}$$

where $\Theta_{j,j-1} = \left(\vec{r}_j, \vec{r}_{j-1}\right)$,
with the angle between two vectors in spherical coordinates being:

$$\cos\Theta_{j,j-1} = \cos\theta_j\cos\theta_{j-1} + \sin\theta_j\sin\theta_{j-1}\cos\left(\varphi_j - \varphi_{j-1}\right), \tag{24}$$

and the measurement takes the form:

$$\prod_{j=1}^{N+1}\left(\frac{m}{2i\pi\hbar\varepsilon}\right)^{\frac{3}{2}}\prod_{j=1}^{N}d\vec{r} = \left(\frac{m}{2i\pi\hbar\varepsilon}\right)^{\frac{3}{2}(N+1)}\prod_{j=1}^{N}\left[\int_0^{\infty}\int_0^{\pi}\int_0^{2\pi} r_j^2\sin\theta_j dr_j d\theta_j d\varphi_j\right]. \tag{25}$$

The previous expression of the propagator is not appropriate for integration due to the presence of the term $\left(\frac{-i}{\hbar}\frac{m}{\varepsilon}r_jr_{j-1}\cos\Theta_{j,j-1}\right)$, and this term is separable into a radial part and an angular part.

For an explicit evolution of the angular part of the propagator, we will use the following formula [6]:

$$\begin{aligned}\exp(iz\cos\varphi) &= 2^{\frac{1}{2}}\Gamma\left(\frac{1}{2}\right)\sum_{k=0}^{+\infty}\left(k+\frac{1}{2}\right)i^k(z)^{\frac{-1}{2}}J_{k+\frac{1}{2}}(z)P_k(\cos\varphi)\\ &= \sqrt{\frac{\pi}{2z}}\sum_{k=0}^{+\infty}(2k+1)i^kJ_{k+\frac{1}{2}}P_k(\cos\varphi).\end{aligned} \tag{26}$$

$J_n(x)$ is the Bessel function, which is given by:

$$J_n(x) = \sum_{P=0}^{+\infty}\frac{(-1)^P}{P!(n+P)!}\left(\frac{x}{2}\right)^{2P+n}, \tag{27}$$

if

$$x = -U, \tag{28}$$

then

$$\begin{aligned}J_n(-U) &= \sum_{P=0}^{+\infty}\frac{(-1)^P}{P!(n+P)!}\left(\frac{-U}{2}\right)^{2P+n}\\ &= \sum_{P=0}^{+\infty}\frac{(-1)^P}{P!(n+P)!}\left(\frac{U}{2}\right)^{2P+n}(-1)^n.\end{aligned} \tag{29}$$

we have

$$J_n(ix) = i^n I_n, \tag{30}$$

where I_n is the modified Bessel function.
We define

$$y = iz, \tag{31}$$

According to (26) and (30), we can deduce that

$$\begin{aligned} e^{y\cos\varphi} &= \sqrt{\frac{-\pi}{2iy}}\sum_{k=0}^{+\infty}(2k+1)i^k J_{k+\frac{1}{2}}P_k(\cos\varphi) \\ &= (-1)^{\frac{1}{2}}\sqrt{\frac{\pi}{2iy}}\sum_{k=0}^{+\infty}(2k+1)i^k(-1)^{k+\frac{1}{2}}J_{k+\frac{1}{2}}(iy)P_k(\cos\varphi) \\ &= (-1)^{\frac{1}{2}}\sqrt{\frac{\pi}{2iy}}\sum_{k=0}^{+\infty}(2k+1)i^k(-1)^{k+\frac{1}{2}}(i)^{k+\frac{1}{2}}I_{k+\frac{1}{2}}(y)P_k(\cos\varphi) \\ &= \sqrt{\frac{\pi}{2y}}\sum_{k=0}^{+\infty}(2k+1)I_{k+\frac{1}{2}}(y)P_k(\cos\varphi), \end{aligned} \tag{32}$$

where $P_k(\cos\varphi)$ are the Legendre polynomials.
We arrive at the following expression for the propagator, by substituting formula (32) in (22):

$$\begin{aligned} K\left(\overrightarrow{r''},t'';\overrightarrow{r'},t'\right) &= \lim_{N\to\infty}\prod_{j=1}^{N+1}\left(\frac{m}{2i\pi\hbar\varepsilon}\right)^{\frac{3}{2}}\prod_{i=1}^{N}\left[\int_0^\infty\int_0^\pi\int_0^{2\pi} r_j^2\sin\theta_j dr_j d\theta_j d\varphi_j\right] \\ &\times \prod_{j=1}^{N+1}\left[\exp\left[\frac{i}{\hbar}\left[\frac{m}{2\varepsilon}\left(r_j^2+r_{j-1}^2\right)-\varepsilon V(r_j)\right]\right]\right] \\ &\times \prod_{j=1}^{N+1}\left(\frac{i\pi\hbar\varepsilon}{2mr_jr_{j-1}}\right)^{\frac{1}{2}}\sum_{l\ldots l_{N+1=0}}^{+\infty}(2l_j+1) \\ &\times I_{l_j+\frac{1}{2}}\left(\frac{mr_jr_{j-1}}{i\hbar\varepsilon}\right)P_l(\cos\Theta_{j,j-1}), \\ &\qquad l_j = l, \forall j = 1, \ldots, N+1 \end{aligned} \tag{33}$$

where else

$$\begin{aligned} K\left(\overrightarrow{r''},t'';\overrightarrow{r'},t'\right) &= \lim_{N\to\infty}\left(\frac{m}{2i\pi\hbar\varepsilon}\right)^{\frac{3}{2}(N+1)}\sum_{l\ldots l_{N+1=0}}^{+\infty}\left[\prod_{j=1}^{N}\int_0^\infty r_j^2 dr_j\right] \\ &\times \left[\prod_{j=1}^{N}\int_0^\pi\int_0^{2\pi}(2l_j+1)I_{l_j+\frac{1}{2}}\left(\frac{mr_jr_{j-1}}{i\hbar\varepsilon}\right)P_{l_j}(\cos\Theta_{j,j-1})\sin\theta_j d\theta_j d\varphi_j\right] \\ &\times \left[(2l_{N+1}+1)I_{l_{N+1}+\frac{1}{2}}\left(\frac{mr_jr_{j-1}}{i\hbar\varepsilon}\right)P_{l_{N+1}}(\cos\theta_{j,j-1})\right] \\ &\times \prod_{j=1}^{N+1}\left[\exp\left[\frac{i}{\hbar}\left[\frac{m}{2\varepsilon}\left(r_j^2+r_{j-1}^2\right)-\varepsilon V(r_j)\right]\right]\right] \\ &\times \prod_{j=1}^{N+1}\left(\frac{i\pi\hbar\varepsilon}{2mr_jr_{j-1}}\right)^{\frac{1}{2}}, \end{aligned} \tag{34}$$

we can use the following expression

$$\prod_{j=1}^{N} r_j = \frac{1}{\sqrt{r'' r'}} \prod_{j=1}^{N+1} \left(r_j r_{j-1}\right)^{\frac{1}{2}}, \tag{35}$$

by substituting Expression (35) in (34), we obtain

$$\begin{aligned} K\left(\overrightarrow{r''}, t''; \overrightarrow{r'}, t'\right) &= \lim_{N\to\infty} \left(\frac{m}{2i\pi\hbar\varepsilon}\right)^{(N+1)} \left(\frac{1}{\sqrt{r'' r'}}\right) \sum_{\ell \ldots \ell_{N+1=0}}^{+\infty} \left[\prod_{j=1}^{N} \int_0^{+\infty} r_j dr_j\right] \\ &\times \prod_{j=1}^{N+1} \left[\exp\left[\frac{i}{\hbar}\left\{\frac{m}{2\varepsilon}\left(r_j^2 + r_{j-1}^2\right) - \varepsilon V(r_j)\right\}\right]\right] \\ &\times \prod_{j=1}^{N} \left[\int_0^{\pi}\int_0^{2\pi} (2\ell_j + 1) I_{\ell_j + \frac{1}{2}}\left(\frac{m r_j r_{j-1}}{i\hbar\varepsilon}\right) P_{\ell_j}\left(\cos\Theta_{j,j-1}\right) \sin\theta_j d_j d\varphi_j\right] \\ &\times \left[\left(2\ell_{N+1} + I_{\ell_{N+1}+\frac{1}{2}}\left(\frac{m r_j r_{j-1}}{i\hbar\varepsilon}\right) P_{\ell_{N+1}}\left(\cos\Theta_{j},_{j-1}\right)\right]. \end{aligned} \tag{36}$$

The Legendre polynomials can be decomposed into spherical harmonics

$$P_\ell(\cos\theta) = \left(\frac{4\pi}{2\ell + 1}\right) \sum_{m=-\ell} Y_{\ell,m}(\theta_N, \varphi_N) Y^*_{\ell,m}(\theta_{N-1}, \varphi_{N-1}), \tag{37}$$

where $P_\ell(\cos\theta)$ represents the Legendre polynomial of order ℓ and degree m, θ is the polar angle, and ϕ is the azimuthal angle. $Y_{\ell,m}(\theta, \varphi)$ corresponds to the associated spherical harmonic of order ℓ and degree m.

This formula establishes a connection between Legendre polynomials and spherical harmonics, providing an expansion in terms of angles for functions or phenomena with spherical symmetry,

where

$$Y_{\ell,m}(\theta, \varphi) = (-1)^m \left[\frac{(2\ell + 1)}{4\pi} \times \frac{(\ell - m)!}{(\ell + m)!}\right] P_\ell^m(\cos\theta) \exp(im\varphi), \tag{38}$$

Formula (32) is as follows:

$$e^{y\cos\varphi} = 2\pi\sqrt{\frac{2\pi}{y}} \sum_{\ell=0}^{+\infty} (2\ell + 1) I_{\ell+\frac{1}{2}}(y) \sum_{m=-\ell}^{\ell} Y_{\ell,m}(\theta_N, \varphi_N) Y^*_{\ell,m}(\theta_{N-1}, \varphi_{N-1}), \tag{39}$$

By inserting the last formula into the propagator expression (36)

$$\begin{aligned}
K\left(\overrightarrow{r''},t'';\overrightarrow{r'},t'\right) &= \lim_{N\to\infty}\left(\frac{m}{i\hbar\varepsilon}\right)^{(N+1)}\left(\frac{1}{\sqrt{r''r'}}\right)\sum_{\ell\ldots\ell_{N+1=0}}^{+\infty}\left[\prod_{j=1}^{N}\int_0^{\infty} r_j dr_j\right] \\
&\times \prod_{j=1}^{N+1}\left[\exp\left[\frac{i}{\hbar}\left\{\frac{m}{2\varepsilon}\left(r_j^2+r_{j-1}^2\right)-\varepsilon V\left(r_j\right)\right\}\right]\right] \\
&\times \prod_{j=1}^{N}\left[I_{\ell_j+\frac{1}{2}}\left(\frac{mr_jr_{j-1}}{i\hbar\varepsilon}\right)I_{\ell_{N+1}+\frac{1}{2}}\left(\frac{mr_jr_{j-1}}{i\hbar\varepsilon}\right)\right] \\
&\times \sum_{m_j=-\ell_j}^{\ell_j}\int_0^{\pi}\int_0^{2\pi} Y_{\ell_j,m_j}\left(\theta_j,\varphi_j\right)Y^*_{\ell_j,m_j}\left(\theta_{j-1},\varphi_{j-1}\right)\sin\theta_j d\theta_j d\varphi_j \\
&\times \sum_{m=-\ell}^{\ell} Y_{\ell_{N+1},m}\left(\theta_{N+1},\varphi_{N+1}\right)Y^*_{\ell_{N+1},m}\left(\theta_N,\varphi_N\right).
\end{aligned} \tag{40}$$

Using the orthogonality relation of spherical harmonics, which is described by the following equation

$$\int_0^{\pi} d\varphi\int_0^{2\pi} Y_{\ell,m}(\theta,\varphi)Y^*_{\ell,m}(\theta,\varphi)\sin\theta d\theta = \delta_{\ell,\ell'},\delta_{m,m'}, \tag{41}$$

thus we find the following expression for the propagator

$$K\left(\overrightarrow{r''},t'';\overrightarrow{r'},t'\right) = \sum_{\ell=0}^{+\infty}\frac{(2\ell+1)}{4\pi}K_\ell(r'',t'';r',t')P_\ell(\cos\Theta), \tag{42}$$

where the radial propagator $K_\ell(r'',t'';r',t')$ is also expressed as

$$\begin{aligned}
K_\ell(r'',t'';r',t') &= \lim_{N\to\infty}\left(\frac{m}{i\hbar\varepsilon}\right)^{(N+1)}\left(\frac{1}{\sqrt{r''r'}}\right)\lim_{N\to\infty}\left[\prod_{j=1}^{N}\int_0^{\infty} r_j dr_j\right] \\
&\times \prod_{j=1}^{N+1}\left[I_{\ell_j+\frac{1}{2}}\left(\frac{mr_jr_{j-1}}{i\hbar\varepsilon}\right)\right] \\
&\times \prod_{j=1}^{N+1}\left[\exp\left[\frac{i}{\hbar}\left\{\frac{m}{2\varepsilon}\left(r_j^2+r_{j-1}^2\right)-\varepsilon V\left(r_j\right)\right\}\right]\right],
\end{aligned} \tag{43}$$

Indeed, considering the asymptotic behavior of the modified Bessel functions, [6].

$$I_j\left(\frac{z}{\varepsilon}\right)\underset{\varepsilon\to 0}{\to}\left(\frac{\varepsilon}{2\pi z}\right)^{\frac{1}{2}}\exp\left\{\frac{z}{\varepsilon}-\frac{1}{2}\frac{\varepsilon}{z}\left(v^2-\frac{1}{4}\right)\right\}, \tag{44}$$

then

$$I_{\ell+\frac{1}{2}}\left(\frac{\frac{mr_jr_{j-1}}{i\hbar}}{\varepsilon}\right)\underset{\varepsilon\to 0}{\to}\left(\frac{\varepsilon i\hbar}{2\pi mr_jr_{j-1}}\right)^{\frac{1}{2}}\exp\left\{\frac{mr_jr_{j-1}}{\varepsilon i\hbar}-\left(\frac{\varepsilon i\hbar}{2mr_jr_{j-1}}\right)(\ell(\ell+1))\right\}, \tag{45}$$

then we arrive at the formulation of the radial propagator in spherical coordinates and as a function of the effective potential $V_{eff}(r_j)$:

$$\begin{aligned} K_\ell(r'',t'';r',t') &= \left(\frac{1}{r''r'}\right) \lim_{N\to\infty} \left(\frac{m}{2\pi i\hbar\varepsilon}\right)^{\frac{(N+1)}{2}} \times \left[\prod_{j=1}^{N} \int_0^\infty dr_j\right] \\ &\times \exp \frac{i}{\hbar} \sum_{j=1}^{N+1} \left\{ \frac{m}{2\varepsilon} (\Delta r_j)^2 - \varepsilon V(r_j) - \frac{\ell(\ell+1)\hbar^2\varepsilon}{2mr_j r_{j-1}} \right\}, \end{aligned} \tag{46}$$

where the effective potential is defined by the following expression:

$$V_{eff}(r_j) = V(r_j) + \frac{\hbar^2 \ell(\ell+1)}{2mr_j r_{j-1}}, \tag{47}$$

So the propagator (46) becomes:

$$\begin{aligned} K_\ell(r'',t'';r',t') &= \left(\frac{1}{r''r'}\right) \lim_{N\to\infty} \left(\frac{m}{2\pi i\hbar\varepsilon}\right)^{\frac{(N+1)}{2}} \times \left[\prod_{j=1}^{N} \int_0^\infty dr_j\right] \\ &\times \exp\left[\frac{i}{\hbar} \sum_{j=1}^{N+1} \left\{ \frac{m}{2\varepsilon} (\Delta r_j)^2 - \varepsilon V_{eff}(r_j) \right\}\right]. \end{aligned} \tag{48}$$

The specific form of the radial propagator will depend on the potential energy term $V(r)$ in the radial Schrödinger equation, which corresponds to the particular physical system being studied. Different potential energy functions will lead to different solutions and, consequently, different forms of the radial propagator.

5. Feynman propagator

The propagator related to a central potential $V(r)$ between two space-time points $\left(\vec{r'},t'\right)$ and $\left(\vec{r''},t''\right)$, in spherical coordinates is written as [4, 5]:

$$K\left(\vec{r''},t'';\vec{r'},t'\right) = \frac{1}{4\pi r'' r'} \sum_{\ell=0}^{\infty} (2\ell+1) \times K_\ell(r'',t'';r',t') P_\ell(\cos\Theta), \tag{49}$$

where $P_\ell(\cos\Theta)$ is the Legendre polynomial and $\Theta \equiv \left(\vec{r''},\vec{r'}\right)$ with

$$K_\ell(r'',t'';r',t') = \lim_{N\to\infty} \int \prod_{j=1}^{N} \exp\left[\frac{i}{\hbar} S_j\right] \times \prod_{j=1}^{N} \left[\frac{m}{2\pi i\hbar\varepsilon}\right]^{\frac{1}{2}} \prod_{j=1}^{N-1} dr_j, \tag{50}$$

where

$$S_j = \frac{m}{2\varepsilon} (\Delta r_j)^2 - \varepsilon V_{eff}(r_j), \tag{51}$$

here

$\Delta r_j = r_j - r_{j-1}, \varepsilon = \Delta t_j = t_j - t_{j-1}, t' = t_0$ and $t'' = t_N$,
and effective potential V_{eff} is defined by the relation as:

$$V_{eff}(r) = \frac{\hbar^2}{2m}\frac{\ell(\ell+1)}{r^2} + V(r), \tag{52}$$

Thus, the condensed form is given by:

$$K_\ell(r'', t''; r', t') = \mathfrak{D}r(t)\exp\left[\int_{t'}^{t''}\left(\frac{m}{2}\dot{r}^2 - V_{eff}(r)\right)dt\right], \tag{53}$$

5.1 Pöschl-Teller potential

This potential is an important diatomic molecular potential. Many applications of the analytical and approximate technique in the current literature have been made to establish eigensolutions and thermodynamic properties [5, 7]. Another example of this potential used as an effective model is as a reference potential manifested to elaborate on the reliability of the order ambiguity parameters.

In the present chapter, the Pöschl-Teller potential of hyperbolic form [5] has been used and is given by:

$$V^{PT}(r) = \left[\frac{A}{\sinh^2(\alpha r)} - \frac{B}{\cosh^2(\alpha r)}\right], \quad r \geq 0, \tag{54}$$

and

$$\begin{cases} A = \dfrac{\hbar^2\alpha^2}{2m}\eta(\eta-1) \\ B = \dfrac{\hbar^2\alpha^2}{2m}\lambda(\lambda+1) \end{cases}, \tag{55}$$

where α, η, λ are positive constants.

5.1.1 s-states ($\ell = 0$)

For $\ell = 0$, taking into account (50), the propagator of the Pöschl-Teller potential (54) becomes:

$$\begin{aligned} K_\ell(r'', r', s) &= \int \mathfrak{D}r(s)\exp\left[\frac{i}{\hbar}\int_0^{s''}\left(\frac{m}{2}\dot{r}^2 - V_{eff}(r)\right)ds\right] \\ &= \int \mathfrak{D}r(s)\exp\left[\frac{i}{\hbar}\int_0^{s''}\left(\frac{m}{2}\dot{r}^2 - \frac{\hbar^2}{2m}\left[\frac{\eta(\eta-1)}{\sinh^2(r)} - \frac{\lambda(\lambda+1)}{\cosh^2(r)}\right]\right)ds\right] \\ &= \int \mathfrak{D}r(s)\mu_{\lambda\,\eta}[\sinh(r)\cosh(r)]\exp\left(\frac{im}{2\hbar}\int_0^{s''}\dot{r}^2 ds\right) \\ &= \lim_{N\to\infty}\left(\frac{m}{2\pi i\varepsilon}\right)^{\frac{N}{2}}\prod_{j=1}^{N-1}\int_0^\infty dr^{(j)}\prod_{j=1}^{N}\mu_{\lambda\,\eta}\left[\sinh(r)^{(j)}\cosh(r)^{(j)}\right]\exp\left(\frac{im}{2\varepsilon\hbar}(r_j - r_{j-1})^2\right), \end{aligned} \tag{56}$$

we use the notation $\widehat{\sinh^2(\theta)}^{(j)} = \sinh(\theta)^{(j)} \sinh(\theta)^{(j-1)}$, when the functional measure $\mu_{\lambda\,\eta}$ given by [4, 5]:

$$\begin{aligned}\mu_{\lambda\,\eta}[\sinh(\alpha r), \cosh(\alpha r)] &= \lim_{N\to\infty} \prod_{j-1}^{N} \mu_{\lambda\,\eta}\left[\sinh(\alpha r)^{(j)}, \cosh(\alpha r)^{(j)}\right] \\ &= \lim_{N\to\infty} \left(\frac{2\pi m}{\varepsilon\hbar}\right)^{N} \prod_{j=1}^{N} \widehat{\sinh(\alpha r)}^{(j)} \widehat{\cosh(\alpha r)}^{(j)} \\ &\times \exp\left(\frac{-m}{i\varepsilon\hbar}\left(\widehat{\sinh^2(\alpha r)}^{(j)} - \widehat{\cosh^2(\alpha r)}^{(j)}\right)\right) \\ &\times I_{\eta-\frac{1}{2}}\left(\frac{m}{i\varepsilon\hbar}\sinh^2(\alpha r)\right) \times I_{\lambda-\frac{1}{2}}\left(\frac{im}{\varepsilon\hbar}\cosh^2(\alpha r)\right),\end{aligned} \tag{57}$$

This is a known solved problem.

Adapting Frank and Wolf's notion, the solution of the path integral reads $2S = \eta(\eta - 1)$, $-2C = \lambda(\lambda + 1)$, and by introducing the numbers k_1, k_2 which are defined as a function of C and S [5], by setting,

$$\begin{cases} k_1 = \frac{1}{2}\left[1 \pm \left(\frac{1}{4} - 2C\right)^{\frac{1}{2}}\right] \\ k_2 = \frac{1}{2}\left[1 \pm \left(\frac{1}{4} + 2S\right)^{\frac{1}{2}}\right] \end{cases}. \tag{58}$$

The propagator $K_\ell(r'', r', T)$ contains discrete and continuous terms, becomes:

$$\begin{aligned}K_\ell(r'', r', T) &= \sum_{n=0}^{N_M} \exp\left(\frac{-is'' E_{n,\ell}^{PT}}{\hbar}\right) \Psi_{\ell,n}^{*(k_1,k_2)}(r') \Psi_{\ell,n}^{(k_1,k_2)}(r'') \\ &+ \int_0^\infty dk \exp\left(-is''\frac{\hbar k^2}{2m}\right) \Psi_k^{*(k_1,k_2)}(r') \Psi_k^{(k_1,k_2)}(r''),\end{aligned} \tag{59}$$

we have N_M indicate the maximum number of states with $0, 1, \ldots, n \le N_M < k_1 - k_2 - \frac{1}{2}$. The signs depend on the boundary conditions for $r \to 0$ and $r \to \infty$, respectively.

The bound states are explicitly given by [4, 5]:

$$\begin{aligned}\Psi_{\ell,n}^{(k_1,k_2)}(r) &= N_n^{(k_1,k_2)} (\sinh(\alpha r))^{2k_2 - \frac{1}{2}} (\cosh(\alpha r))^{-2k_1 + \frac{3}{2}} \\ &\times_2 F_1\left(-k_1 + k_2 + k, -k_1 + k_2 - k + 1; 2k_2; -\sinh^2(\alpha r)\right) \\ &= \left(\frac{2n!(2k_1 - 1)\Gamma(2k_1 - n - 1)}{\Gamma(2k_2 + n)\Gamma(2k_1 - 2k_2 - n)}\right)^{\frac{1}{2}} (\sinh(\alpha r))^{2k_2 - \frac{1}{2}} (\cosh(\alpha r))^{2n - 2k_1 + \frac{3}{2}} \\ &\times P_n^{[2k_2 - 1, 2(k_1 - k_2 - n) - 1]}\left(\frac{1 - \sinh^2(\alpha r)}{\cosh^2(\alpha r)}\right),\end{aligned} \tag{60}$$

and

$$N_n^{(k_1,k_2)} = \frac{1}{\Gamma(2k_2)}\left(\frac{(2k-1)\Gamma(k_1+k_2-k)\Gamma(k_1+k_2+k-1)}{\Gamma(k_1-k_2+k)\Gamma(k_1-k_2-k+1)}\right). \tag{61}$$

the energy spectrum is also obtained by:

$$E_n^{PT} = -\left(\frac{\hbar^2\alpha^2}{2m}\right)(2k-1)^2 = -\left(\frac{\hbar^2\alpha^2}{2m}\right)[2(k_1-k_2-n)-1]^2. \tag{62}$$

5.1.2 ℓ-states ($\ell \neq 0$)

Usually, we find that the effective potential is not exactly solvable for ℓ-states ("$\ell \neq 0$"), To deal with the centrifugal term $\left(\frac{1}{r^2}\right)$, we need to find a better approximate expression for this term and such approximations have been proposed as a general approximation similar to the type of Pöschl-Teller potential [8]:

$$\frac{1}{r^2} \approx F(r) = \alpha^2\left(\frac{1}{3\cosh^2(\alpha r)} + \frac{1}{\sinh^2(\alpha r)}\right), \tag{63}$$

Moreover, these approximations are only valid for small values of the parameter α and collapse for large α. This choice is useful and allows us to treat this hyperbolic potential.

Substituting (63) into (52) we find:

$$V_{eff}(r) = \frac{\hbar^2\alpha^2}{2m}\left[\frac{\eta_1(\eta_1-1)}{\sinh^2(\alpha r)} - \frac{\lambda_1(\lambda_1+1)}{\cosh^2(\alpha r)}\right] + C_1, \tag{64}$$

with

$$\begin{cases} \eta_1(\eta_1-1) = \left[\frac{2\hbar^2\alpha^2}{m}d_1\left[\left(\ell+\frac{D}{2}-1\right)^2-\frac{1}{4}\right]+\eta(\eta-1)\right] \\ \lambda_1(\lambda_1+1) = \left[-\frac{\hbar^2\alpha^2}{m}d_0\left[\left(\ell+\frac{D}{2}-1\right)^2-\frac{1}{4}\right]+\lambda(\lambda+1)\right], \\ C_1 = \frac{2\hbar^2\alpha^2}{m}d_2\left[\left(\ell+\frac{D}{2}-1\right)^2-\frac{1}{4}\right] \end{cases} \tag{65}$$

with the bound states being explicitly given by [5]:

$$\begin{aligned} \Psi_{\ell,n}^{(k_1,k_2)}(r) &= N_n^{(k_1,k_2)}(\sinh(\alpha r))^{2k_2-\frac{1}{2}}(\cosh(\alpha r))^{-2k_1+\frac{3}{2}} \\ &\times_2 F_1(-k_1+k_2+k, -k_1+k_2-k+1; 2k_2; -\sinh^2(\alpha r)) \\ &= \left(\frac{2n!(2k_1-1)\Gamma(2k_1-n-1)}{\Gamma(2k_2+n)\Gamma(2k_1-2k_2-n)}\right)^{\frac{1}{2}}(\sinh(\alpha r))^{2k_2-\frac{1}{2}}(\cosh(\alpha r))^{2n-2k_1+\frac{3}{2}} \\ &\times P_n^{[2k_2-1,2(k_1-k_2-n)-1]}\left(\frac{1-\sinh^2(\alpha r)}{\cosh^2(\alpha r)}\right), \end{aligned} \tag{66}$$

and

$$N_n^{(k_1,k_2)} = \frac{1}{\Gamma(2k_2)}\left(\frac{(2k-1)\Gamma(k_1+k_2-k)\Gamma(k_1+k_2+k-1)}{\Gamma(k_1-k_2+k)\Gamma(k_1-k_2-k+1)}\right). \tag{67}$$

the energy spectrum is also obtained by:

$$E_{n,\ell}^{PT} = -\left(\frac{\hbar^2\alpha^2}{2m}\right)(2k-1)^2 = -\left(\frac{\hbar^2\alpha^2}{2m}\right)[2(k_1-k_2-n)-1]^2. \tag{68}$$

with

$$\begin{cases} k_1 = \frac{1}{2}\left[1 \pm \left(\frac{1}{4}+\lambda_1(\lambda_1+1)\right)^{\frac{1}{2}}\right] \\ k_2 = \frac{1}{2}\left[1 \pm \left(\frac{1}{4}+\eta_1(\eta_1-1)\right)^{\frac{1}{2}}\right] \end{cases}. \tag{69}$$

The energy spectrum is obtained from Eq. (69), namely

$$E_{n,\ell}^{PT} = -\left(\frac{\hbar^2\alpha^2}{2m}\right)\left[2\left(\begin{array}{c} \frac{1}{2}\left[1 \pm \left(\frac{1}{4}+\left(\frac{2m}{\hbar^2\alpha^2}\left[-\frac{\hbar^2\alpha^2}{6m}\frac{\left[\left(l+\frac{D}{2}-1\right)^2-\frac{1}{4}\right]}{3}+\frac{B}{q}\right]\right)\right)^{\frac{1}{2}}\right] \\ -\frac{1}{2}\left[1 \pm \left(\frac{1}{4}+\left(\frac{2m}{\hbar^2\alpha^2}\left[\frac{\hbar^2\alpha^2}{2m}\left[\left(l+\frac{D}{2}-1\right)^2-\frac{1}{4}\right]+\frac{A}{q}\right]\right)\right)^{\frac{1}{2}}\right]-n \end{array}\right)-1\right]^2, \tag{70}$$

6. Duru-Kleinert method

We often introduce a coordinate transformation followed by a local time transformation to make the study much more accessible.

Let us perform the following space and time changes [9]:

$$\begin{cases} r = f(q) \\ dt = f'^2(q)ds \end{cases}, \tag{71}$$

These transformations allow us to transform a difficult propagator to calculate into a more manageable form.

Moreover, Green's function relative to a given propagator allows us to derive from its poles the spectrum of energies and the corresponding wave functions from the residues at the poles. This function is obtained from the Fourier transform of the propagator $K_\ell(r'', r'; T)$ as follows:

$$K_\ell(r'', r'; T) = \frac{1}{2\pi\hbar}\int G_\ell(r'', r'; E)\exp\left(\frac{-iET}{\hbar}\right)dE, \tag{72}$$

with

$$G_\ell(r'',r';E) = \frac{i}{\hbar}\left[f'(q')f'(q'')\right]^{\frac{1}{2}} \int_0^\infty \hat{K}_\ell(q'',q';s'')ds'', \tag{73}$$

and

$$\hat{K}_\ell(q'',q';s'') = \int_{q'}^{q''} \mathfrak{D}q(s) \exp\left[\frac{i}{\hbar}\int_0^{s''}\left(\frac{m}{2}\dot{q}^2 - f'^2(q)\left[V_{eff}(q) - E\right] - \Delta V(q)\right)ds\right], \tag{74}$$

and the quantum correction ΔV is given by:

$$\Delta V(q) = \frac{\hbar^2}{8m}\left[3\frac{\left[f''(q)\right]^2}{\left[f'(q)\right]^2} - 2\frac{f'''(q)}{f'(q)}\right] \tag{75}$$

7. Energy spectrum and wave functions

7.1 Shifted Deng-Fan Oscillator potential

Another important empirical potential of diatomic molecules is the Shifted Deng-Fan Oscillator potential [7]. It was proposed since more than half century ago, but has attracted much interest lately, and this potential is the form

$$V_{SDF}(r) = D_1\left(1 - \frac{b}{e^{\alpha r} - 1}\right)^2 - D_2, \quad b = e^{\alpha r_e} - 1, \tag{76}$$

where D_2 is the dissociation energy, r_e is the position of the minimum, and α denotes the radius of the potential.

Here, we use for this potential a different approximation obtained using a power series decomposition [10, 11].

$$\frac{1}{r^2} \simeq \frac{1}{r_e^2}\left[C_0 + \frac{C_1}{e^{\alpha r} - 1} + \frac{C_2}{\left(e^{\alpha r} - 1\right)^2}\right], \tag{77}$$

where r_e is the minimum of the potential (76) and

$$\begin{cases} C_0 = \dfrac{\left(1 - \dfrac{(1-\eta)^2}{u^2}\left(\dfrac{4u}{1-\eta} - (3+u)\right)\right)}{u^2} \\ C_1 = \dfrac{\left(\exp(u)(1-\eta)^2\right)}{u^3} \\ C_2 = \dfrac{\left(\exp(2u)(1-\eta)^4\right)}{u^4}\left(3 + u - \dfrac{2u}{1-\eta}\right) \end{cases},$$

where $u = 2\alpha r_e$, and $\eta = \exp(-u)$.
Substituting Eqs. (77) and (78) into Eq. (52), we find

$$V_{eff}(r) = \frac{\hbar^2}{2m}\left[C_0 + \frac{C_1 \exp(-2\alpha r)}{1 - \exp(-2\alpha r)} + \frac{C_2 \exp(-2\alpha r)}{(1 - \exp(-2\alpha r))^2}\right] + D_1\left(1 - \frac{b}{\exp(\alpha r) - 1}\right)^2 - D_2, \tag{78}$$

In *a* more compact for me, it reads

$$V_{eff}(r) = -A\coth(\alpha r) + \frac{B}{\sinh^2(\alpha r)} + C, \tag{79}$$

where

$$\begin{cases} A = \frac{\hbar^2}{2m}\ell(\ell+1)\alpha^2\left(\frac{C_2}{2} - \frac{C_1}{2}\right) + D_1 b + D_1\frac{b^2}{2} \\ B = \frac{\hbar^2}{2m}\ell(\ell+1)\alpha^2\frac{C_2}{4} + D_1\frac{b^2}{4} \\ C = \frac{\hbar^2}{2m}\ell(\ell+1)\alpha^2\left(C_0 - \frac{C_1}{2} + \frac{C_2}{2}\right) + D_1 + D_1 b + D_1\frac{b^2}{2} - D_2 \end{cases}, \tag{80}$$

Thus, the condensed form is given by:

$$K_\ell(r'', t''; r', t') = \mathfrak{D}r(t)\exp\left[\int_{t'}^{t''}\left(\frac{m}{2}\dot{r}^2 - V_{eff}(r)\right)dt\right], \tag{81}$$

the potential given by (79) is similar to the Manning-Rosen, a direct path integration is not possible, the problem can be solved with the help of the folowing space-time transformation

$$\begin{cases} r = F(q) = \frac{1}{\alpha}arccoth\left(2\coth^2(q) - 1\right) \\ dt = [F'(q)]^2 ds \end{cases}, \tag{82}$$

According to [7], the wave function is given by

$$\begin{aligned} \chi_{n,\ell}^{S.D.F(k_1,k_2)}(r) &= \sqrt{\alpha}N_n^{(k_1,k_2)}(1-u)^{1/2-k_1+n}(u)^{k_1-1-s/2-n} \\ &\quad \times {}_2F_1\left(-n, 2k_1 - n - 1; s+1; \frac{1}{1-u}\right) \\ &= \left[\frac{\alpha(2k_1-1)n!\Gamma(2k_1-n-1)}{\Gamma(n+s+1)\Gamma(2k_1-s-n-1)}\right]^{1/2}\left(1-e^{-2r}\right)^{k_2}\exp\left[-2r\left(k_1 - \frac{s}{2} - n - 1\right)\right] \\ &\quad \times P_n^{(2k_2-2n-s-2,s)}\left(1 - 2e^{-2r}\right), \end{aligned} \tag{83}$$

where $P_n^{(\alpha,\beta)}$ denotes the Jacobi polynomials and $u = \frac{1}{2}[1 - \tanh(2\alpha r)]$, where $k = k_1 - k_2 - n$.
and

$$k_1 = \frac{1}{2}\left[\left(1 + \frac{1}{2}(s + 2n + 1)\right) + \frac{2mA}{\alpha^2\hbar^2(s + 2n + 1)}\right], \tag{84}$$

$$k_2 = \frac{1}{2}\left[1 + \sqrt{1 + \frac{8mB}{\alpha^2\hbar^2}}\right] \equiv \frac{1}{2}(1 + s), \tag{85}$$

The energy spectrum is obtained from the poles of the Green function, Eq. (82), namely

$$E_{n,l}^{SDF} = -\left[\frac{\alpha^2\hbar^2(s + 2n + 1)^2}{8m} + \frac{2mA^2}{\alpha^2\hbar^2(s + 2n + 1)^2}\right] + C. \tag{86}$$

7.2 Generalized inverse quadratic Yukawa potential

The generalized inverse quadratic Yukawa potential extends this concept by introducing additional parameters or modifications to the potential. These modifications can include terms that account for different types of interactions or other physical phenomena, depending on the specific context or application.

The general form of Generalized Inverse Quadratic Yukawa Potential is:

$$V_{GIQY}(r) = -a - b\frac{e^{-\alpha r}}{r} - c\frac{e^{-2\alpha r}}{r^2}, \tag{87}$$

which means that the effective potential becomes

$$V_{eff}(r) = -a - b\frac{e^{-\alpha r}}{r} - c\frac{e^{-2\alpha r}}{r^2} + \frac{\hbar^2\ell(\ell + 1)}{2r^2}. \tag{88}$$

First of all, we deal with the centrifugal terms using the approximation [10, 11].

$$\frac{1}{r^2} = \frac{4\alpha^2 e^{-2\alpha r}}{(1 - e^{-2\alpha r})^2}, \tag{89}$$

and

$$\frac{1}{r} = \frac{2\alpha e^{-\alpha r}}{1 - e^{-2\alpha r}}, \tag{90}$$

putting these considerations together, we find the following:

$$V_{eff}(r) = -b\frac{2\alpha e^{-2\alpha r}}{1 - e^{-2\alpha r}} - c\frac{4\alpha^2 e^{-4\alpha r}}{(1 - e^{-2\alpha r})^2} + \frac{\hbar^2\ell(\ell + 1)}{2}\frac{4\alpha^2 e^{-2\alpha r}}{(1 - e^{-2\alpha r})^2} - a, \tag{91}$$

$V_{eff}(r)$ can be reformulated as

$$V_{eff}(r) = A\coth(\alpha r) + \frac{B}{\sinh^2(\alpha r)} + C, \tag{92}$$

with

$$\begin{cases} A = \alpha(2c\alpha - b); \\ B = \alpha^2\left(\dfrac{\hbar^2}{2m}\ell(\ell+1) - c\right); \\ C = -2c\alpha^2 + b\alpha - a. \end{cases} \tag{93}$$

Since the difficulties of doing the integration of Eq. (53) straightforwardly, we perform a space-time transformation depending on the Duru-Kleinert method [4, 9], so we do a nontrivial change of variable $(r \to q)$ followed by time local transformation $(t \to s)$

$$\begin{cases} r = h(q) = \dfrac{1}{\alpha}\text{argcoth}\left(2\coth^2(q) - 1\right); \\ t \to s \Leftrightarrow dt = \left[h'(q(s))\right]^2 ds. \end{cases} \tag{94}$$

Putting these considerations together, we find the new Green's function

$$G_\ell(q_b, q_a; E) = \frac{i}{\hbar}\left[h'(q_a)h'(q_b)\right]^{\frac{1}{2}}\int_0^\infty P_\ell(q_b, q_a; S)dS, \tag{95}$$

where h' is the derivative of h with respect to q, and the new form of the promotor is

$$P_\ell(q_b, q_a; S) = \int Dq(s)\exp\left[\frac{i}{\hbar}\int_0^S\left\{\frac{m}{2}\dot{q}^2 - h'^2\left(V_{eff}(q) - E\right) - \Delta V(q)\right\}ds\right], \tag{96}$$

the quantum correction $\Delta V(q)$ [4] is given by

$$\Delta V(q) = \frac{\hbar^2}{8m}\left(3\frac{h''^2}{h'^2} - 2\frac{h'''}{h'}\right) = \frac{\hbar^2}{8m}\left(\frac{1}{\cosh^2(q)} + \frac{1}{\sinh^2(q)}\right), \tag{97}$$

and the transformed effective potential is

$$V_{eff}(q) = A\left(2\coth^2(q) - 1\right) + 2B\left(2\coth^2(q) - 2\right)\coth^2(q) + C, \tag{98}$$

therefore

$$\begin{aligned} h'^2\left(V_{eff}(q) - E\right) + \Delta V(q) &= \frac{\hbar^2}{2m}\left(\frac{\frac{8mB}{\alpha^2\hbar^2} + \frac{3}{4}}{\sinh^2(q)} + \frac{\frac{2m}{\alpha^2\hbar^2}(E + A - C) + \frac{1}{4}}{\cosh^2(q)}\right) \\ &- \frac{1}{\alpha^2}(E - A - C). \end{aligned} \tag{99}$$

And using the following abbreviations

$$\begin{cases} D = \dfrac{1}{\alpha^2}(E - A - C); \\ \eta^2 - \dfrac{1}{4} = \dfrac{8mB}{\alpha^2\hbar^2} + \dfrac{3}{4}; \\ \upsilon^2 - \dfrac{1}{4} = -\dfrac{2m}{\alpha^2\hbar^2}(E + A - C) - \dfrac{1}{4}, \end{cases} \tag{100}$$

which means that

$$\begin{cases} D = \frac{1}{\alpha^2}(E - A - C); \\ \eta = \pm\sqrt{1 + \frac{8mB}{\alpha^2\hbar^2}}; \\ \upsilon = \pm\sqrt{-\frac{2m}{\alpha^2\hbar^2}(E + A - C)}, \end{cases} \tag{101}$$

we can rewrite the promotor as follows:

$$\begin{aligned} P_\ell(q_b, q_a; s) &= \int Dq(s) \exp\left[\frac{i}{\hbar}\int_0^S \left\{\frac{m}{2}\dot{q}^2 - \frac{\hbar^2}{2m}\left(\frac{\eta^2 - \frac{1}{4}}{\sinh^2(q)} + \frac{\upsilon^2 - \frac{1}{4}}{\cosh^2(q)}\right)\right\} ds\right] \\ &\times \exp\left[\frac{i}{\hbar}DS\right], \end{aligned} \tag{102}$$

which is nothing but a promotor formula corresponding to a system with modified Pöschl-Teller potential and energy D [12], and accordingly, the integration over time S enables us to obtain directly the radial Green's function related to this system

$$G_\ell(q_b, q_a; D) = \int_0^\infty P_\ell(q_b, q_a; S)dS, \tag{103}$$

thus

$$\begin{aligned} G_\ell(q_b, q_a; D) &= \int_0^\infty dS \exp\left[\frac{i}{\hbar}DS\right] \\ &\times \int Dq(s) \exp\left[\frac{i}{\hbar}\int_0^S \left\{\frac{m}{2}\dot{q}^2 - \frac{\hbar^2}{2m}\left(\frac{\eta^2 - \frac{1}{4}}{\sinh^2(q)} + \frac{\upsilon^2 - \frac{1}{4}}{\cosh^2(q)}\right)\right\} ds\right], \end{aligned} \tag{104}$$

The energy spectrum is obtained from the poles of Green's function which leads us to

$$D = -\frac{\hbar^2}{2m}(2n + \eta - \upsilon + 1)^2, \tag{105}$$

therefore

$$\begin{aligned} E_{n,\ell} &= -\frac{\hbar^2\alpha^2}{8m}\left(2n + 1 + \sqrt{1 + \frac{8mB}{\alpha^2\hbar^2}}\right)^2 \\ &- \frac{2A^2}{\frac{\hbar^2\alpha^2}{m}\left(2n + 1 + \sqrt{1 + \frac{8mB}{\alpha^2\hbar^2}}\right)^2} + C, \end{aligned}$$

the energy spectrum is thus

$$\begin{aligned} E_{n,\ell} &= -\frac{\hbar^2\alpha^2}{8m}\left(2n+1+\sqrt{1+4\ell(\ell+1)-\frac{8m}{\hbar^2}c}\right)^2 \\ &- \frac{2(2c\alpha-b)^2}{\frac{\hbar^2}{m}\left(2n+1+\sqrt{1+4\ell(\ell+1)-\frac{8m}{\hbar^2}c}\right)^2} - 2c\alpha^2 + b\alpha - a, \end{aligned} \tag{106}$$

On the other hand, the associated wave functions can be displayed as

$$\begin{aligned} \psi_n(r) &= \left[\left(\alpha - \frac{4mA}{\alpha\hbar^2(\omega+2n+1)^2}\right)\frac{(2k_1-2n-\omega-2)n!\Gamma(2k_1-n-1)}{\Gamma(n+\omega+1)\Gamma(2k_1-\omega-n-1)}\right]^{1/2} \\ &\times (1-\exp(-2\alpha r))^{\frac{\omega+1}{2}}\exp(k_1-\omega/2-n-1) \\ &\times P_n^{(2k_1-2n-\omega-2,\omega)}(1-2\exp(-2\alpha r)), \end{aligned}$$

where $P_n^{(2k_1-2n-\omega-2,\omega)}$ are Jacobi polynomials with the notations

$$\begin{cases} k_1 = \frac{1}{2}\left(1+\frac{1}{2}(\omega+2n+1)-\frac{2mA}{\alpha^2\hbar^2(\omega+2n+1)}\right); \\ k_2 = \frac{1}{2}(1+\omega), \end{cases} \tag{107}$$

and

$$\omega = \sqrt{1+\frac{8mB}{\alpha^2\hbar^2}}. \tag{108}$$

7.3 Modified screened Coulomb plus inversely quadratic Yukawa potential

The Modified Screened Coulomb plus Inversely Quadratic Yukawa potential (MSC-IQY) is a combined potential energy function that incorporates both the screened Coulomb potential and the inversely quadratic Yukawa potential. This modified potential is often used in various areas of physics to describe interactions between charged particles, taking into account both screening effects and long-range Coulombic interactions. For $a = 0$, the GIQY potential reduces to Modified Screened Coulomb Plus Inversely Quadratic Yukawa potential (MSC-IQY) of the form

$$V_{GIQY}(r) = -b\frac{e^{-\alpha r}}{r} - c\frac{e^{-2\alpha r}}{r^2}, \tag{109}$$

and the associated energy eigenvalues are obtained as

$$\begin{aligned} E_{n,\ell}^{GIQY} &= -\frac{\hbar^2\alpha^2}{8m}\left(2n+1+\sqrt{1+4\ell(\ell+1)-\frac{8m}{\hbar^2}c}\right)^2 \\ &- \frac{2(2c\alpha-b)^2}{\frac{\hbar^2}{m}\left(2n+1+\sqrt{1+4\ell(\ell+1)-\frac{8m}{\hbar^2}c}\right)^2} - 2c\alpha^2 + b\alpha. \end{aligned} \tag{110}$$

7.4 Kratzer potential

The Kratzer potential [13] is a mathematical model used to describe the interaction between a particle and a central force field. It is commonly employed to study molecular systems and the vibrational motion of diatomic molecules. For $\alpha = 0, a = 0, b = 2D_e r_e$, and $c = D_e r_e^2$, the GIQY potential (87) reduces to the Kratzer potential of the form

$$V_K(r) = -2D_e\left(\frac{r_e}{r} - \frac{1}{2}\frac{r_e^2}{r^2}\right), \tag{111}$$

where r_e is the equilibrium bond length and D_e is the dissociation energy. The energy eigenvalues of the Kratzer potential are obtained as

$$E^K_{n,\ell} = -\frac{b^2}{\frac{\hbar^2}{2m}\left(2n+1+\sqrt{1+4\ell(\ell+1)-\frac{8m}{\hbar^2}c}\right)^2}, \tag{112}$$

thus

$$E^K_{n,\ell} = -\frac{(2D_e r_e)^2}{\frac{\hbar^2}{2m}\left(2n+1+\sqrt{1+4\ell(\ell+1)-\frac{8m}{\hbar^2}D_e r_e^2}\right)^2}. \tag{113}$$

7.5 Yukawa potential

The Yukawa potential, also known as the screened Coulomb potential or the Debye-Hückel potential, is a mathematical model used to describe the interaction between charged particles with an exponential decay due to screening effects. It is commonly employed in physics to study phenomena such as electromagnetic interactions, nuclear forces, and scattering processes.

The Yukawa potential is given by the following equation (setting $a = 0$ and $c = 0$, Eq. (88) takes the form)

$$V_Y(r) = -b\frac{e^{-\alpha r}}{r}, \tag{114}$$

which is known as Yukawa potential, its corresponding energy eigenvalues achieved are

$$\begin{aligned} E^Y_{n,\ell} &= -\frac{\hbar^2\alpha^2}{8m}\left(2n+1+\sqrt{1+4\ell(\ell+1)}\right)^2 \\ &\quad - \frac{2b^2}{\frac{\hbar^2}{m}\left(2n+1+\sqrt{1+4\ell(\ell+1)}\right)^2} + b\alpha, \end{aligned} \tag{115}$$

or equivalently

$$E^Y_{n,\ell} = -\frac{\hbar^2\alpha^2}{2m}(n+1+\ell)^2 - \frac{b^2}{\frac{\hbar^2}{m}2(n+1+\ell)^2} + b\alpha, \tag{116}$$

7.6 Inversely quadratic Yukawa potential

The Inversely Quadratic Yukawa potential (IQY) is a modified version of the Yukawa potential that takes into account an additional inverse square term. As $a = 0$ and $b = 0$, Eq. (88) reduces to the Inversely Quadratic Yukawa potential (IQY) of the form

$$V_{IQY}(r) = -c\frac{e^{-2\alpha r}}{r^2}, \tag{117}$$

the energy eigenvalue equation becomes

$$\begin{aligned} E_{n,\ell}^{IQY} &= -\frac{\hbar^2\alpha^2}{8m}\left(2n+1+\sqrt{1+4\ell(\ell+1)-\frac{8m}{\hbar^2}c}\right)^2 \\ &- \frac{2(2c\alpha)^2}{\frac{\hbar^2}{m}\left(2n+1+\sqrt{1+4\ell(\ell+1)-\frac{8m}{\hbar^2}c}\right)^2} - 2c\alpha^2. \end{aligned} \tag{118}$$

7.7 Coulomb potential

The Coulomb potential is used to calculate important properties such as the electric potential, electric field, and electrostatic forces in systems involving charged particles. It forms the basis for understanding phenomena such as the behavior of ions in solutions, the interaction between charged particles in plasmas, and the structure of atoms and molecules.

When $a = 0, \alpha = 0$, and $c = 0$, Eq. (88) reduces to the Coulomb potential of the form

$$V_C(r) = -\frac{b}{r}, \tag{119}$$

the energy eigenvalues of the Coulomb potential are obtained as

$$E_{n,\ell}^{C} = -\frac{b^2}{\frac{\hbar^2}{2m}\left(2n+1+\sqrt{1+4\ell(\ell+1)}\right)^2}, \tag{120}$$

hence

$$E_{n,\ell}^{C} = -\frac{2m}{\hbar^2}\frac{b^2}{2(n+\ell+1)^2}, \tag{121}$$

8. Conclusions

We have presented a rigorous treatment using the path integral approach of Feynman. We affirm that this formalism is an efficient and powerful tool for finding the propagator associated with several problems in quantum physics, particularly nonrelativistic problems. Most of these problems cannot be treated exactly, and practically no physical system can be studied without approximation methods.

In this chapter, we have adopted a two-step approach to study exponentially shaped potentials. In the first step, by introducing a judicious approximation to handle the centrifugal term, we were able to transition from solving a problem related to ℓ-states to that of the s-state. The other step involves adapting a spatio-temporal transformation by Duru-Kleinert. The use of this transformation was revisited in the reasoning process to reduce the unsolvable relative propagator to the effective potential of several potentials, specifically to the modified Pöschl-Teller potential. This problem is well known and was previously addressed within the framework of the Schrödinger formulation and the path integral. The energy spectrum and wave functions are determined in this case.

In conclusion, our method is effective in solving this type of potential. We hope to continue developing the path integral formalism not only for exponential-type potentials but also for other types and more general forms, and in other domains of physics.

Acknowledgements

The authors would like to thank the **LESI** laboratory of the University of Khemis Miliana for its help in carrying out this study.

Author details

Hocine Boukabcha[1*†], Salah Eddin Aid[2†] and Amina Ghobrini[3]

1 Laboratory of Energy and Intelligent Systems, University of Khemis Miliana, Khemis Miliana, Algeria

2 Laboratory of Mechanics and Energy, Hassiba Benbouali University of Chlef, Chlef, Algeria

3 Laboratory of Coatings, Materials and Environment University Mhamed Bougara of Boumerdes, Boumerdes, Algeria

*Address all correspondence to: h.boukabcha@univ-dbkm.dz

† These authors contributed equally.

References

[1] Messiah A. Mécanique Quantique. Paris: Dunod; 1964

[2] Landau L, Lifchitz E. Mécanique Quantique. Tome III. Editions Mir: Moscou; 1967

[3] Ince P. Ordinary differential equations. New York: Dover publications INC; 1966

[4] Kleinert H. Path integrals in Quantum Mechanics, Statistics, Polymer Physics and Financial Markets. World Scientific: Singapore; 2009

[5] Grosche C. Path integral solution of a class of potentials related to the Pöschl-Teller potential. Journal of Physics A: Mathematical and General. 1989;**22**:5073. DOI: 10.1088/0305-4470/22/23/012

[6] Nikiforov AF, Uvarov VB. Special Functions of Mathematical Physics. Basel: Birkhauser; 1988

[7] Boukabcha H, Hachama M, Diaf A. Ro-vibrational energies of the shifted Deng-Fan oscillator potential with Feynman path intcgral formalism. Applied Mathematics and Computation. 2018;**321**:121-129

[8] Badawi R, Bessis N, Bessis G. On the introduction of the rotation-vibration coupling in diatomic molecules and the factorization method. Journal of Physics B: Atomic and Molecular Physics. 1972;**5**: L157-L161. DOI: 10.1088/0022-3700/5/8/004

[9] Duru IH, Kleinert H. Solution of the path integral for the H-atom. 1979;**84**: 185. DOI: 10.1016/0370-2693(79)90280-6

[10] Greene RL, Aldrich C. Variational wave functions for a screened Coulomb potential. Physical Review A. 1976;**14**: 2363. DOI: 10.1103/PhysRevA.14.2363

[11] Gönül Bözer O, Canelik Y, Koak M. Hamiltonian hierarchy and the Hulthèn potential. Physics Letters A. 2000;**275**: 238-243

[12] Aid SE, Boukabcha H, Benzaid D. Non-relativistic treatment of generalized inverse quadratic Yukawa potential via path integral approach. Indian Journal of Physics. 2023;**97**(7):1989-1995

[13] Kratzer A. Die ultraroten rotations spektren der halogenwasserstoffe. Zeitschrift fr Physik. 1920;**3**:289-307

Chapter 4

The Inverse of the Discrete Momentum Operator

Armando Martínez-Pérez and Gabino Torres-Vega

Abstract

In the search of a quantum momentum operator with discrete spectrum, we obtain some properties of the discrete momentum operator for nonequally spaced spectrum. We find the inverse operator. We use the matrix representation of these operators, and we find that there is one more eigenvalue and eigenfunction than the dimension of the matrix. We apply the results to obtain the discrete adjoint of the momentum operator. We conclude that we can have discrete operators which can be self-adjoint and that it is possible to define a self-adjoint extension of the corresponding Hilbert space. These results help us understand the quantum time operator.

Keywords: discrete quantum mechanics, discrete momentum operator, inverse of the momentum operator, nonstandard finite differences derivative, exact discrete integration

1. Introduction

Nonstandard finite difference derivatives help determine the discrete versions of some differential equations and their solutions [1–10]. This method uses nonstandard expressions of the finite differences derivative in such a way that they give the exact result when applied to a particular function.

Another benefit of nonstandard finite difference for the derivative of a function is that it can be used as a discrete quantum operator to deal with quantum mechanical operators with discrete spectrum [11, 12]. Since some quantum operators have a discrete spectrum, a discrete derivative can be very useful in quantum mechanics theory [11, 12].

In Section 2, we define and obtain some properties of the discrete derivative operator from a global point of view, i.e., considering all the values of a function on all the points of a mesh at once. This is done by defining a matrix that collects the derivatives for each mesh point when applied to a given vector. We find the eigenvalues and eigenvectors of the derivative matrix. We also discuss the commutation properties between the derivative and coordinate matrices. The canonical commutator is satisfied only along some directions.

The summation by parts theorem and the adjoint of the momentum operator are found in Section 3. We introduce the discrete symmetric operator definition similar to continuous variables functions in a Hilbert space.

An interesting result is that the considered matrices have more eigenvalues and corresponding *almost* eigenvectors (the last entry of the eigenvector is null) beside the usual number due to their dimension of them. For a semi-infinite matrix, the last entry is of little effect, and such additional eigenvalues will belong to the matrix spectrum when seen as an operator. Such additional eigenvalues and eigenvectors are common to all the considered matrices. With these results, we can say that there are also self-adjoint discrete operators and that we can also have discrete self-adjoint extensions in the corresponding Hilbert space. These results are beneficial when dealing with the question of the existence of a time operator in quantum mechanics [12].

We introduce the discrete inverse matrix of the discrete derivative operator in Section 4. The difference between the scheme we address in this work with other proposals for a discrete derivative is a modification in the derivative matrix for the final point of a grid of points, which causes the derivative matrix to have an inverse.

We can deal with any mesh without asking for equidistant points. At the end of this paper, there are some concluding remarks.

2. Discrete derivation

Let us consider a partition $\mathcal{P} = \{q_0, q_1, q_2, \dots, q_N\}$ of the interval $[q_0, q_N]$ and vectors $\mathbf{f} = (f_0, f_1, \dots, f_N)^T$, and $\mathbf{g} = (g_0, g_1, \dots, g_N)^T$ associated to this partition. The distances $\Delta_j = q_{j+1} - q_j$, for each j, are not supposed to be equal.

The finite differences derivative matrix $\mathbf{D}$ is defined as

$$\mathbf{D} = \begin{pmatrix} -\frac{1}{\xi_0} & \frac{1}{\xi_0} & 0 & \dots & 0 & 0 \\ 0 & -\frac{1}{\xi_1} & \frac{1}{\xi_1} & \dots & 0 & 0 \\ 0 & 0 & -\frac{1}{\xi_2} & \dots & 0 & 0 \\ \vdots & & & & & \\ 0 & 0 & 0 & \dots & -\frac{1}{\xi_{N-1}} & \frac{1}{\xi_{N-1}} \\ 0 & 0 & 0 & \dots & 0 & -\frac{1}{\xi_N} \end{pmatrix}, \tag{1}$$

where

$$\xi_j = \Delta_j e^{-ip\Delta_j/2} \mathrm{sinc}\left(\frac{\Delta_j}{2} p\right), \quad j = 0, \dots, N-1, \tag{2}$$

$$\xi_N = -\frac{i}{p}. \tag{3}$$

The function sinc(z) is the entire function equal to one at $z = 0$ and $z^{-1} \sin z$ otherwise. The continuous parameter p in this expression is related to the conjugate variable to the discrete variable q_j, see Eq. (11) below. The choice of ξ_j ensures that the finite differences derivative (d-derivative) delivers the exact result when acting on the complex exponential function e^{-ipq}.

In case it is needed, for small Δ_j we have the power series expansion

$$\xi_j \approx \Delta_j - i\frac{p}{2}\Delta_j^2 - \frac{p^2}{6}\Delta_j^3, \quad 0 \leq j < N. \tag{4}$$

We see that ξ_j is similar to the difference Δ_j of the usual finite differences derivative. However, we will only consider the case where Δ_j has a finite value.

Let us discuss some properties of the d-derivative matrix. The action of the d-derivative matrix $\mathbf{D}$ when acting to the left, on the vector $\mathbf{f}^T = (f_0, f_1, \dots, f_N)$, results in

$$\mathbf{f}^T\mathbf{D} = \left(-\frac{f_0}{\xi_0}, -(\mathcal{D}\mathbf{f})_1, -(\mathcal{D}\mathbf{f})_2, \dots, -(\mathcal{D}\mathbf{f})_N\right), \tag{5}$$

where

$$(\mathcal{D}\mathbf{f})_j = \frac{f_j}{\xi_j} - \frac{f_{j-1}}{\xi_{j-1}}, \tag{6}$$

is a finite differences approximation to the derivative of a function extended to the complex plane. These improved increments ξ_j are defined over the complex plane. For a small difference Δ_j, we have that

$$(\mathcal{D}\mathbf{f})_j \approx \left(\frac{f_{j+1}}{\Delta_{j+1}} - \frac{f_j}{\Delta_j}\right) + i\frac{p}{2}\left(f_{j+1} - f_j\right) + \frac{p^2}{12}\left(f_{j+1} - f_{j+1}\right)\Delta_{j+1}. \tag{7}$$

We see that we have another discrete approximation to the derivative of a function.

Now, the action to the right of the derivative matrix on a vector is:

$$\mathbf{D}\mathbf{g} = \left((\mathrm{D}g)_0, (\mathrm{D}g)_1, \dots, (\mathrm{D}g)_{N-1}, -\frac{f_N}{\xi_N}\right)^T, \tag{8}$$

where

$$(\mathrm{D}g)_j = \frac{g_{j+1} - g_j}{\xi_j}, \tag{9}$$

is a modified finite differences derivative of $g(q)$ at q_j. In case Δ_j is small, we have that

$$(\mathbf{D}\mathbf{g})_j \approx \frac{g_{j+1} - g_j}{\Delta_j} + i\frac{p}{2}\left(g_{j+1} - g_j\right) - \frac{p^2}{12}\Delta_j\left(g_{j+1} - g_j\right). \tag{10}$$

The first term in this approximation is the usual finite differences derivative of a function.

Note that for in the limiting case, $\Delta_j \to 0$, both nonstandard finite differences (Eq. 5) and (Eq. 8) reduce to the usual forward finite difference approximation to the derivative.

The eigenvalues of the derivative matrix **D** are $\lambda_j = -1/\xi_N, -1/\xi_{N-1}, \dots, -1/\xi_0$, and the corresponding eigenvectors are

$$\left\{ \begin{pmatrix} \dfrac{\xi_N^N}{\prod_{n=0}^{N-1}(\xi_N - \xi_n)} \\ \dfrac{\xi_N^{N-1}}{\prod_{n=1}^{N-1}(\xi_N - \xi_n)} \\ \dfrac{\xi_N^{N-2}}{\prod_{n=2}^{N-1}(\xi_N - \xi_n)} \\ \vdots \\ 1 \end{pmatrix}, \begin{pmatrix} \dfrac{\xi_{N-1}^{N-1}}{\prod_{n=0}^{N-2}(\xi_{N-1} - \xi_n)} \\ \dfrac{\xi_{N-1}^{N-2}}{\prod_{n=1}^{N-2}(\xi_{N-1} - \xi_n)} \\ \dfrac{\xi_{N-1}^{N-3}}{\prod_{n=2}^{N-3}(\xi_{N-1} - \xi_n)} \\ \vdots \\ 0 \end{pmatrix}, \dots, \begin{pmatrix} 1 \\ 0 \\ 0 \\ \vdots \\ 0 \end{pmatrix} \right\}. \tag{11}$$

Note that, due to the operator character of the matrix, there is an additional eigenvector, the exponential function $\mathbf{e} = \left(e^{-ipq_0}, e^{-ipq_1}, e^{-ipq_2}, \dots, e^{-ipq_N}\right)$, with eigenvalue $-ip$,

$$\mathbf{De} = -ip\,\mathbf{e}. \tag{12}$$

3. The adjoint of the discrete derivative

A sesquilinear form between vectors **f** and **g** is defined with the help of the summation matrix:

$$\mathbf{S} = \begin{pmatrix} \xi_0 & 0 & 0 & \dots & 0 & 0 \\ 0 & \xi_1 & 0 & \dots & 0 & 0 \\ 0 & 0 & \xi_2 & \dots & 0 & 0 \\ \vdots & & & & & \\ 0 & 0 & 0 & \dots & \xi_{N-1} & 0 \\ 0 & 0 & 0 & \dots & 0 & 0 \end{pmatrix}, \tag{13}$$

obtaining

$$\begin{aligned} &\mathbf{f}^T\mathbf{SDg} \\ &= -f_0g_0 + f_0g_1 - f_1g_1 + f_1g_2 - f_2g_2 + f_2g_3 - f_3g_3 + \dots + f_Ng_N \\ &= \mathbf{g}^T\left(\mathbf{B} - \mathbf{S}\tilde{\mathbf{D}}\right)\mathbf{f}, \end{aligned} \tag{14}$$

where

$$\tilde{\mathbf{D}} = \begin{pmatrix} 0 & 0 & 0 & 0 & 0 & 0 \\ -\dfrac{1}{\xi_1} & \dfrac{1}{\xi_1} & 0 & \dots & 0 & 0 \\ 0 & -\dfrac{1}{\xi_2} & \dfrac{1}{\xi_2} & \dots & 0 & 0 \\ \vdots & & & & & \\ 0 & 0 & 0 & \dots & \dfrac{1}{\xi_{N-1}} & 0 \\ 0 & 0 & 0 & \dots & 0 & 0 \end{pmatrix}. \tag{15}$$

and

$$\mathbf{B} = \begin{pmatrix} -1 & 0 & 0 & 0 & 0 & 0 \\ 0 & 0 & 0 & \dots & 0 & 0 \\ 0 & 0 & 0 & \dots & 0 & 0 \\ \vdots & & & & & \\ 0 & 0 & 0 & \dots & 0 & 0 \\ 0 & 0 & 0 & \dots & 1 & 0 \end{pmatrix}. \tag{16}$$

Eq. (13) is the summation by parts equality in matrix form. We call the matrix $\tilde{\mathbf{D}}$ the d-adjoint of the discrete derivative matrix $\mathbf{D}$.

A row of the summation by parts matrix equality is:

$$\sum_{n=0}^{N-1} \xi_n f_n (\mathbf{Dg})_n + \sum_{n=1}^{N-1} \xi_n (\tilde{\mathbf{D}}\mathbf{f})_n g_n = f_{N-1} g_N - f_0 g_0, \tag{17}$$

which is the discrete version of the integration by parts theorem of the calculus of continuous variables.

The previous results are useful in quantum mechanics theory when considering the momentum or the Hamiltonian operators with a discrete spectrum.

We define the discrete momentum operator at q_j as

$$\hat{P}_j = -i(\mathrm{D})_j, \quad 0 \le j < N, \tag{18}$$

and its adjoint

$$\hat{P}_j^{\dagger} = -i(\tilde{\mathrm{D}})_j, \quad 0 < j < N. \tag{19}$$

The summation by parts provides the adjoint of the momentum operator and its symmetry property. Explicitly, Eq. (16) is rewritten as

$$\sum_{n=0}^{N-1} \xi_n f_n^* (-i\mathbf{Dg})_n - \sum_{n=1}^{N} \xi_n \left[\left(-i\tilde{\mathbf{D}}^* \mathbf{f}\right)_n\right]^* g_n = -if_{N-1}^* g_N + if_0^* g_0, \tag{20}$$

This equality yields

$$\langle f|\hat{P}g\rangle = \langle \hat{P}f|g\rangle = -if_{N-1}^* g_N + if_0^* g_0. \tag{21}$$

Thus, we say that the discrete momentum operator $\hat{P}$ is d-symmetric, if $f_{N-1} = e^{i\theta} f_0$ and $g_N = e^{i\theta} g_0$, as is the case for continuous variables operators.

It is also possible to consider self-adjoint extensions for the discrete momentum operator, as it is done for the case of the continuous variable momentum operator [13].

3.1 Commutator between the d-derivative and the coordinate

In general, a discrete canonical commutation relationship $[A, B] = I$ is not possible for finite-dimensional matrices A and B because the trace of this relationship results in a contradiction ((0 = 1) [14]. However, there are some directions in which the commutator evaluates to a constant different from zero: the directions pointed at by its eigenvectors, for example. In addition, the matrix can be considered as an operation with additional eigenfunctions.

If we call $\mathbf{Q} = \text{diag}\left(q_j\right)$ to the coordinate matrix, the usual commutator between the d-derivative and coordinate matrices is:

$$[\mathbf{D}, \mathbf{Q}] = \begin{pmatrix} 0 & \frac{\Delta_0}{\xi_0} & 0 & \dots & 0 & \dots & 0 \\ 0 & 0 & \frac{\Delta_1}{\xi_1} & \dots & 0 & \dots & 0 \\ 0 & 0 & 0 & \dots & 0 & \dots & 0 \\ \vdots & \vdots & \vdots & \ddots & \vdots & \ddots & \vdots \\ 0 & 0 & 0 & \dots & 0 & \dots & \frac{\Delta_{N-1}}{\xi_{N-1}} \\ 0 & 0 & 0 & \dots & 0 & \dots & 0 \end{pmatrix}. \tag{22}$$

This matrix shifts and rescales the vector entries on which it acts. This matrix approaches an identity matrix when $\Delta_j \to 0$.

For a finite Δ_j, we look for the eigenvectors of the commutator matrix to obtain a diagonal matrix. The eigenvalues of the commutator (21), considered as a matrix, are all zero with multiplicity $N+1$. The eigenvectors are $(1, 0, \dots, 0)^T$ and $(0, \dots, 0)^T$ with multiplicity N. In addition to considering the eigenvectors of this commutator matrix to obtain a diagonal matrix, we can take advantage of rescaling to cancel shifting and return to the original vector. Then, the commutator matrix (21) has the additional eigenvector

$$\mathbf{h}^T = \left(\frac{1}{\lambda^{N-1}} \prod_{j=0}^{N-2} \frac{\Delta_j}{\xi_j}, \frac{1}{\lambda^{N-2}} \prod_{j=1}^{N-2} \frac{\Delta_j}{\xi_j}, \dots, \frac{\lambda \xi_{N-1}}{\Delta_{N-1}}\right)^T, \tag{23}$$

with an eigenvalue λ. The action of the commutator matrix on these vectors results in the same vector with the last entry equal to zero, which is almost an eigenvector.

Still another eigenvector, with eigenvalue one, is

$$\tilde{\mathbf{h}}^T = \left(1, \frac{\xi_0}{\Delta_0}, \dots, \prod_{n=0}^{N-1} \frac{\xi_n}{\Delta_n}\right)^T, \tag{24}$$

The commutator is equal to one along this direction. Then, the canonical commutation relationship is also valid in this direction.

Thus, along the mentioned directions, the d-derivative has similar properties as its continuous variable counterpart.

4. The inverse of the d-derivative

The d-derivative matrix that we use can be inverted. The determinant of the d-derivative matrix is

$$|\mathbf{D}| = \frac{1}{\xi_0\xi_1\xi_2\xi_3 \dots \xi_N}. \tag{25}$$

The inverse of the d-derivative matrix **D** is the negative of the progressive discrete integration matrix

$$\mathbf{I} = \begin{pmatrix} \xi_0 & \xi_1 & \xi_2 & \xi_3 & \dots & \xi_N \\ 0 & \xi_1 & \xi_2 & \xi_3 & \dots & \xi_N \\ 0 & 0 & \xi_2 & \xi_3 & \dots & \xi_N \\ 0 & 0 & 0 & \xi_3 & \dots & \xi_N \\ \vdots & & & & & \\ 0 & 0 & 0 & 0 & \dots & 0 \end{pmatrix}. \tag{26}$$

We discuss some properties of the d-integration matrix **I**. When the d-integration matrix **I** is applied to the left to a vector $\mathbf{f}^T$ results in

$$\mathbf{f}^T\mathbf{I} = (\mathcal{I}_0\mathbf{f}, \mathcal{I}_1\mathbf{f}, \mathcal{I}_2\mathbf{f}, \dots, \mathcal{I}_N\mathbf{f}), \tag{27}$$

where

$$\mathcal{I}_j\mathbf{f} = \xi_j\left(f_0 + f_1 + \dots + f_j\right), \quad j \le N. \tag{28}$$

The entries of the resulting vector are the progressive discrete integrations of **f** when the subintervals are of equal length ξ_j. When the d-integration matrix is applied to the right, we get

$$\mathbf{Ig} = \left(\mathrm{I}_0\mathbf{g}, \mathrm{I}_1\mathbf{g}, \mathrm{I}_2\mathbf{g}, \dots, \mathrm{I}_N\mathbf{g}\right), \tag{29}$$

where

$$\mathrm{I}_j\mathbf{g} = g_j\xi_j + g_{j+1}\xi_{j+1} + \dots + g_N\xi_N, \quad 0 \le j \le N. \tag{30}$$

This result is the progressive discrete integration of **g** when the subintervals are of different lengths.

The eigenvalues of **I** are $\xi_0, \dots, \xi_N$, and its eigenvectors are the same as for **D**, Eq. (10). But, there is the additional eigenvector $\mathbf{e} = \left(e^{-ipq_0}, \dots, e^{-ipq_N}\right)^T$,

$$\mathbf{Ie} = \frac{i}{p}\mathbf{e}. \tag{31}$$

The d-derivative and its inverse are constant along the same directions. The domain of the d-derivative and d-integration is the same.

Now, the commutator between **S** and **Q** is

$$[\mathbf{Q}, \mathbf{I}] = \begin{pmatrix} 0 & \xi_1(q_1 - q_0) & \xi_2(q_2 - q_0) & \xi_3(q_3 - q_0) & \dots & -\xi_N(q_N - q_0) \\ 0 & 0 & \xi_2(q_2 - q_1) & \xi_3(q_3 - q_1) & \dots & -\xi_N(q_N - q_1) \\ 0 & 0 & 0 & \xi_3(q_3 - q_2) & \dots & -\xi_N(q_N - q_2) \\ \vdots & & & & & \\ 0 & 0 & 0 & 0 & \dots & -\xi_N(q_N - q_{N-1}) \\ 0 & 0 & 0 & 0 & \dots & 0 \end{pmatrix}, \tag{32}$$

which is the progressive discrete integral of $g(q)\left(q - q_j\right)$ when acting on the vector **g**.

5. Conclusions

We have found another property of the d-derivative matrix: its inverse. The inverse of the d-derivative has the right properties; the properties of the continuous variable integration.

We discussed some of the properties of the discrete momentum operator when considering all of a subset of the spectrum points at once and its associated discrete integration matrices. The matrices are related by a common eigenvector for continuous variable functions. These results give us confidence that our choice is a good candidate for the discrete quantum momentum operator.

We also found that the matrices associated with the discrete derivative and the discrete integration have an additional eigenvalue and eigenvector, in contrast with the usual behavior of standard matrices. We have increased the number of eigenvalues and eigenvectors of a matrix by using it as an operator.

These operators are of help in defining a time operator and its eigenvalues and eigenvectors for use in nonrelativistic quantum mechanics [12]. They can also be used when the angular momentum on a circle is considered [15–17].

These results imply that we can deal with discrete quantum operators in almost the same way as for continuous variable operators case, including deficiency indices and self-adjoint extensions [13].

We have considered the exact discrete derivative for the complex exponential function, but these results are also valid for the real exponential function e^{-pq} with the replacements

$$\xi_N = \frac{1}{p}, \tag{33}$$

$$\xi_j = \frac{1 - e^{-p\Delta_j}}{p}, \quad j < N. \tag{34}$$

Acknowledgements

A. Martínez-Pérez would like to acknowledge the support from the UNAM Postdoctoral Program (POSDOC).

Author details

Armando Martínez-Pérez[1†] and Gabino Torres-Vega[2*†]

1 IIMAS, UNAM, México City, México

2 Departamento de Fisica, Cinvestav, México City, México

*Address all correspondence to: gabino.torres@cinvestav.mx; gabino@fis.cinvestav.mx

† These authors contributed equally.

References

[1] Mickens RE. Nonstandard finite difference models of differential equations. Singapore: World Scientific; 1994

[2] Mickens M. Discretizations of nonlinear differential equations using explicit nonstandard methods. Journal of Computational and Applied Mathematics. 1999;**110**:181

[3] Mickens RE. Nonstandard finite difference schemes for differential equations. The Journal of Difference Equations and Applications. 2010;**8**:823

[4] Mickens RE. Applications of Nonstandard Finite Difference Schemes. Singapore: World Scientific; 2000

[5] Mickens RE. Calculation of denominator functions for nonstandard finite difference schemes for differential equations satisfying a positivity condition. Numerical Methods for Partial Differential Equations. 2006;**23**:672

[6] Potts RB. Differential and difference equations. The American Mathematical Monthly. 1982;**89**:402-407

[7] Potts RB. Ordinary and partial differences equations. Journal of the Australian Mathematical Society: Series B, Applied Mathematics. 1986;**27**:488

[8] Tarasov VE. Exact discretization by Fourier transforms. Communications in Nonlinear Science and Numerical Simulation. 2016;**37**:31

[9] Tarasov VE. Exact discrete analogs of derivatives of integer orders: differences as infinite series. Journal of Mathematics. 2015;**2015**:134842. DOI: 10.1155/2015/134842

[10] Tarasov VE. Exact discretization of Schrödinger equation. Physics Letters A. 2016;**380**:68. DOI: 10.1016/j.physleta.2015.10.039

[11] Martínez Pérez A, Torres-Vega G. Exact finite differences. The derivative on non uniformly spaced partitions. Symmetry. 2017;**9**:217. DOI: 10.3390/sym9100217

[12] Martínez-Pérez A, Torres-Vega G. Discrete self-adjointness and quantum dynamics. Travel times. J Math Phys. 2021;**62**:012013. DOI: 10.1063/5.0021565

[13] Gitman DM, Tyutin IV, Voronov BL. Self-adjoint Extensions in Quantum Mechanics. General Theory and Applications to Schrödinger and Dirac Equations with Singular Potentials. New York: Birkhäuser; 2012

[14] Putnam CR. Commutation Properties of Hilbert Space Operators and Related Topics. Berlin: Springer-Verlag; 1967

[15] Kowalski K, Ławniczak K. Wigner functions and coherent states for the quantum mechanics on a circle. Journal of Physics A: Mathematical and Theoretical. 2021;**54**:275302. DOI: 10.1088/1751-8121/ac019d

[16] Řeháček J, Bouchal Z, Čelechovský R, Hradil Z, Sánchez-Soto LL. Experimental test of uncertainty relations for quantum mechanics on a circle. Physical Review A. 2008;**77**:032110. DOI: 10.1103/PhysRevA.77.032110

[17] Bahr B, Liegener K. Towards exploring features of Hamiltonian renormalisation relevant for quantum gravity. Classical and Quantum Gravity. 2022;**39**:075010. DOI: 10.1088/1361-6382/ac5050

Section 2

Schrödinger Equation – Potential Applications

Chapter 5

Perspective Chapter: Relativistic Treatment of Spinless Particles Subject to a Class of Multiparameter Exponential-Type Potentials

José Juan Peña, Jesús Morales and Jesús García-Ravelo

Abstract

By using the exactly-solvable Schrödinger equation for a class of multi-parameter exponential-type potential, the analytical bound state solutions of the Klein-Gordon equation are presented. The proposal is based on the fact that the Klein-Gordon equation can be reduced to a Schrödinger-type equation when the Lorentz-scalar and vector potential are equal. The proposal has the advantage of avoiding the use of a specialized method to solve the Klein-Gordon equation for a specific exponential potential due that it can be derived by means of an appropriate choice of the involved parameters. For this, to show the usefulness of the method, the relativistic treatment of spinless particles subject to some already published exponential potentials are directly deduced and given as examples. So, beyond the particular cases considered in this work, this approach can be used to solve the Klein-Gordon equation for new exponential-type potentials having hypergeometric eigenfunctions. Also, it can be easily adapted to other approximations of the centrifugal term different to the Green-Aldrich used in this work.

Keywords: Schrödinger-type equation, Klein-Gordon equation, exponential-type potentials, Greene-Aldrich approximation, hypergeometric equation

1. Introduction

At high energy levels, the study of physical phenomena is carried out by means of equations invariant under Lorentz transformations. That is, it requires relativistic wave equations that may be used as a starting point to evaluate the spin-orbit interactions and relativistic effective core potential in the Schrödinger Hamiltonian. The energy levels from these calculations are aimed to find the positions of experimental spectral lines and to predict lines not heretofore observed in the systems under consideration. To that purpose, the Dirac and Klein-Gordon equations are used for the dynamic description of particles with and without spin, respectively. For that, the solutions of these equations have been an important field of research by employing several methods as well as different physical potential models; usually a solution method for a specific potential. In this regard, exponential-type potentials are significant in the study of various physical systems, particularly for modeling diatomic molecules. Within the different exponential potential models, stand out the proposals of Hulthén, Eckart, Manning-Rosen, Rosen-Morse, Deng-Fan, Hyllerass, etc., as well as mixed models with two or more of the above potential models. These latter, have been developed with the aim to solve, as particular cases, the specific potentials that are involved. In any case, the common feature of any exponential-type potential is that wave functions are of hypergeometric-type. For this reason, in the quantum mechanics treatment of this kind of potentials, the method that is most often used to find the bound states solutions is the Nikiforov-Uvarov method [1], which is based on solving a hypergeometric-type differential equation (DE) by means of special orthogonal functions. Albeit, other procedures such as Asymptotic Iteration [2], Supersymmetric Quantum Mechanics [3], He's Variational iteration [4], large-N solutions [5] or Quantization-rule [6], among many other methods, have been also employed in both non-relativistic and relativistic studies; obviously, including numerical solutions [7]. In the relativistic studies of spinless particles, it is well known that the Klein-Gordon equation [8, 9] can always be reduced to a Schrödinger-type equation when the Lorentz-scalar and vector potential are equal [10]. This fact is used in the present research, devoted to obtaining approximate bound state energy eigenvalues and the corresponding eigenfunctions of the Klein-Gordon equation for exponential-type potentials. The method is based on a direct approach applied to the exactly solvable Schrödinger equation with hypergeometric solutions for exponential-type potential [11], which is given in Section 2. With this result, the corresponding analytical bound state solutions of the Klein-Gordon equation are found in the frame of the Green and Aldrich approximation to the centrifugal term [12] as shown in Section 4. Advantageously, according to the method, several specific potentials are derived as particular cases from the proposal such as those given in Section 5.

2. Direct approach to the exactly solvable Schrödinger equation with hypergeometric solutions

For finding exactly-solvable quantum exponential-type potentials, the Schrödinger equation must be transformed into a hypergeometric differential equation. To do so, let us consider the Schrödinger equation ($\hbar^2 = 2m = 1$)

$$\frac{-d^2\psi(r)}{dr^2} + V(r)\psi(r) = E\psi(r) \tag{1}$$

such that, after using the transformation

$$\psi(r) = e^{-\alpha r}u(r) \tag{2}$$

it is written as

$$\frac{-d^2u(r)}{dr^2} + 2\alpha\frac{du(r)}{dr} + \left[V(r) - \left(E + \alpha^2\right)\right]u(r) = 0 \tag{3}$$

With the aim of relating the above equation with a hypergeometric DE, we use the coordinate transformation

$$x = qe^{-r/k}, k > 0, q \neq 0 \tag{4}$$

such that Eq. (3) is written as

$$x(1-x)\frac{d^2u(x)}{dx^2} - (1+2\alpha k)(1-x)\frac{du(x)}{dx} - k^2\left(\frac{1-x}{x}\right)\left[V(x) - \left(E+\alpha^2\right)\right]u(x) = 0. \tag{5}$$

Then, the similarity transformation

$$u(x)(1-x)^d F(x) \tag{6}$$

with d being a real parameter, gives rise to

$$x(1-x)F''(x) + [(1+2\alpha k) - (1+2\alpha k+2d)x]F'(x) + R(x)F(x) = 0 \tag{7}$$

where

$$R(x) = \frac{xd(d-1)}{1-x} - d(1+2\alpha k) - k^2\left(\frac{1-x}{x}\right)\left[V(x) - \left(E+\alpha^2\right)\right]; \tag{8}$$

Hence, Eq. (7) can be compared with a hypergeometric DE

$$x(1 \quad x)y''(x) + [c - (1+a+b)x]y'(x) - aby(x) = 0 \tag{9}$$

provided that

$$1 + 2\alpha k = c, 2(k\alpha + d) = a + b, E = -\alpha^2 \tag{10}$$

and

$$R(x) = -ab \wedge F(x) = y(x) = {}_2F_1(a, b; c : x), \tag{11}$$

where ${}_2F_1$ is the hypergeometric function.
So, from $R = ab$ and $E = -\alpha^2$ in Eq. (8), it is possible to identify the potential

$$V(x) = \frac{1}{k^2}\left[\frac{(ab - (1+2k\alpha)d)x}{1-x} + \frac{d(d-1)x^2}{(1-x)^2}\right] \tag{12}$$

which, by using Eqs. (4) and (10), can be written as

$$V(r) = \frac{[4ab - 2c(a+b+1-c)]qe^{-r/k} + \left[(a-b)^2 - (c-1)^2\right]q^2e^{-2r/k}}{4k^2\left(1-qe^{-r/k}\right)^2} \tag{13}$$

with eigenfunction given from Eqs. (2) and (6) by

$$\psi(r) = e^{-\alpha r}\left(1 - qe^{-r/k}\right)^{d} {}_2F_1\left(a, b; c : qe^{-r/k}\right) \tag{14}$$

and eigenvalue

$$E = -\alpha^2 = -\left(\frac{c-1}{2k}\right)^2. \tag{15}$$

At this point, it is convenient to introduce the new parameters A, B, and C such that $4k^2(A+B) = 4ab - 2c(a+b+1-c)$ and $4k^2(C-A) = (a-b)^2 - (c-1)^2$ in order to rewrite the potential as a multi-parameter exponential-type potential

$$V(r) = \frac{Aqe^{-r/k}}{1 - qe^{-r/k}} + \frac{Bqe^{-r/k}}{(1 - qe^{-r/k})^2} + \frac{Cq^2e^{-2r/k}}{(1 - qe^{-r/k})^2} \tag{16}$$

By solving the condition $\frac{d}{dr}V(r) = 0$, there will be a minimum value for the potential with sufficient depth for the existence of bound states, namely

$$V(r_{min}) = \frac{-(A+B)^2}{4(B+C)}$$
$$with r_{min} = kln\left(\frac{q(A-B-2C)}{A+B}\right), \tag{17}$$

provided that

$$A + B < 0 \leq B + C \tag{18}$$

which ensures that $V(r)$ is an attractive potential with an infinite wall at its singular point $r_s = kln(q)$.

Regarding the wave function, in order to have a node at r_s it is necessary to apply the condition $\psi(r_s) = 0$, which is achieved if $d > 0$. Furthermore, by combining the identities $4k^2(A+B)$ and $4k^2(C-A)$ given above, we have

$$b = \frac{(h+1)(h-a) - 2k^2(A+B)}{h+1-2a}, c = \frac{2a(h-a) - 2k^2(A+B)}{h+1-2a} \tag{19}$$

such that

$$c = a + b - h \, with \, h = \sqrt{1 + 4k^2(B+C)} \tag{20}$$

besides, if the parameter $a = -n, n = 0, 1, 2, 3 \ldots .,$ the hypergeometric function appearing in the eigenfunction given by Eq. (14) becomes a polynomial of $n-th$ degree in the variable $qe^{-r/k}$. Additionally, the condition $\psi(r) \to 0$ when $r \to \infty$ implies that $\alpha = \frac{c-1}{k} > 0$, from which, the number of states is

$$0 \leq n < k\sqrt{C-A} - \frac{h+1}{2} \tag{21}$$

In short, the above equations lead to a well define wave function for a legitimate Schrödinger equation with potential $V(r)$. Likewise, by using Eqs. (19) and (20) in Eq. (15) the energy spectrum will be

$$E_n = \frac{-1}{4k^2}\left[\frac{\left(n+\frac{(h+1)}{2}\right)^2 + k^2(A-C)}{n+\frac{(h+1)}{2}}\right]^2 \tag{22}$$

with wave-functions

$$\psi_n = \left(e^{-r/k}\right)^{\frac{c-1}{2}}\left(1-qe^{-r/k}\right)^{\frac{h+1}{2}} {}_2F_1\left(-n, b; c : qe^{-r/k}\right) \tag{23}$$

3. Klein-Gordon equation in arbitrary dimensions

For a spinless particle with energy E_{nl} and mass M, the D-dimensional Klein-Gordon (KG) equation is given ($\hbar = c = 1$) by [13].

$$-\nabla_D^2 \psi_{nlm}(r,\Omega) + \left\{[M+S(r)]^2 - [E_{nl}+V(r)]^2\right\}\psi_{nlm}(r,\Omega) = 0 \tag{24}$$

where $V(r)$ and $S(r)$ are respectively the Lorentz vector and the scalar interaction potentials. The D-dimensional Laplacian operator in the space $(r,\Omega) = (r, \theta_1, \theta_2, \theta_3, \ldots \theta_{D-2}, \phi)$ is defined ase

$$\nabla_D^2 = r^{1-D}\frac{\partial}{\partial r}\left[r^{D-1}\frac{\partial}{\partial r}\right] - \frac{\Lambda^2(\Omega)}{r^2} \tag{25}$$

with $\Lambda(\Omega)$ the angular momentum operator. Hence, the functione

$$\psi_{nlm}(r,\Omega) = R_{nl}(r)Y_l^m(\Omega) \tag{26}$$

leads to the radial part of Eq. (24)

$$\begin{aligned} &\frac{-d^2R_{nl}(r)}{dr^2} - \frac{D-1}{r}\frac{dR_{nl}(r)}{dr} + \frac{l_D(l_D+1)}{r^2}R_{nl}(r) \\ &+\left\{[M+S(r)]^2 - [E_{nl}+V(r)]^2\right\}R_{nl}(r) = ER_{nl}(r) \end{aligned} \tag{27}$$

where we have used $l_D = \sqrt{\frac{1}{4} + l(l+D-2)} - \frac{1}{2}$ such that

$$\Lambda^2(\Omega)Y_l^m(\Omega) = l_D(l_D+1)Y_l^m(\Omega) \tag{28}$$

Likewise, with $R_{nl}(r) = r^{\frac{-D-1}{2}}\psi_{nL}(r)$ Eq. (27) becomes

$$\frac{-d^2\psi_{nL}(r)}{dr^2} + \left\{[M+S(r)]^2 - [E_{nL}+V(r)]^2 + \frac{L_s}{r^2}\right\}\psi_{nL}(r) = E_{nL}\psi_{nL}(r) \tag{29}$$

where $L_s = L(L+1)$, $L = l_D + \frac{D-3}{2}$ such that the case $D = 3$ implies $L = l_D = l$.

At this point, as already mentioned, the D-dimensional KG equation given in Eq. (29) can be reduced to a Schrödinger-like equation, provided that the Lorentz vector and scalar potential are equal [10]. In fact, if $V(r) = S(r)$ the corresponding KG equation is given by

$$\frac{-d^2\psi_{nL}(r)}{dr^2} + \left\{(E_{nL}^2 - M^2) - 2(E_{nL} + M)V(r) + \frac{L_s}{r^2}\right\}\psi_{nL}(r) = 0 \tag{30}$$

Different methods of the solution have been applied for solving the above equation with many models of interaction potentials; see for example Ikhdair [14] and references therein. Hence, to provide a unified treatment to the bound states solution of the KG equation for equal vector and scalar exponential-type potentials, in the next paragraph the results of Section 2 are extended to consider the D-dimensional case. As we will see, this is done by a simple redefinition of the parameters that appear in $V(r)$ as defined in Eq. (16).

4. Klein-Gordon equation for exponential-type potentials in arbitrary dimensions

To deal with the ℓ-state approximate solutions for the D-dimensional KG equation with the multi-parameter exponential-type potential given in Eq. (16) we define $q = 1$ and

$$A = A + \frac{\alpha L_s}{k^2}, B = B + \frac{\beta L_s}{k^2},\ C = C + \frac{\gamma L_s}{k^2} \tag{31}$$

such that

$$V(r) = \frac{Ae^{-r/k}}{1 - e^{-r/k}} + \frac{Be^{-r/k}}{(1 - e^{-r/k})^2} + \frac{Ce^{-2r/k}}{(1 - e^{-r/k})^2} + L_s T_c \tag{32}$$

with

$$T_c = \frac{(\alpha + \beta)e^{-r/k} + (\gamma - \alpha)e^{-2r/k}}{k^2(1 - e^{-r/k})^2} \tag{33}$$

where, according to the values of the parameters α, β, and γ, the function T_c would be approximate to the centrifugal term. Consequently, this method accepts different approximation schemes, such as recently shown in [15]. For example, if we consider the case $\alpha = \gamma = 0$, and $\beta = 1$, it leads to the standard Green-Aldrich approximation [12]

$$T_c = \frac{e^{-r/k}}{k^2(1 - e^{-r/k})^2} \approx \frac{1}{r^2} \tag{34}$$

however, if we add the constant $c_0 L/k^2$ in both sides of the Eq. (32) one has

$$V(r) + \frac{c_0 L}{k^2} = \frac{Ae^{-r/k}}{1 - e^{-r/k}} + \frac{Be^{-r/k}}{(1 - e^{-r/k})^2} + \frac{Cqe^{-2r/k}}{(1 - e^{-r/k})^2} + L_s\left[T_c + \frac{c_0}{k^2}\right] \tag{35}$$

which means that any improvement to the centrifugal term through an additive constant will be reflected as an additional term in the energy spectrum. In fact, the improved Green-Aldrich approximation to the centrifugal term is achieved when $\alpha = \gamma = 0$, $\beta = 1$ and $c_0 = 1/12$, that is [16]

$$T_c + \frac{c_0}{k^2} = \frac{e^{-r/k}}{k^2\left(1 - e^{-r/k}\right)^2} + \frac{1}{12k^2} \approx \frac{1}{r^2} \tag{36}$$

Another typical improved approximation used to $\frac{1}{r^2}$ is when $\alpha = C_1$; $\beta = 0$, and $\gamma = C_2$ and $c_0 = C_0$ where the parameters C_1, C_2, and C_0 are adjustable parameters [17], leading to

$$T_c + \frac{c_0}{k^2} = \frac{1}{k^2}\left[C_0 + \frac{C_1 e^{-r/k}}{1 - e^{-r/k}} + \frac{C_2 e^{-2r/k}}{\left(1 - e^{-r/k}\right)^2}\right] \approx \frac{1}{r^2} \tag{37}$$

However, for the sake of simplicity, we will use the standard Green-Aldrich approximation, $C_0 = 0$, such that

$$V(r) = V(r) + \frac{L_s}{r^2} \tag{38}$$

with

$$V(r) = \frac{Ae^{-r/k}}{1 - e^{-r/k}} + \frac{Be^{-r/k}}{\left(1 - e^{-r/k}\right)^2} + \frac{Ce^{-2r/k}}{\left(1 - e^{-r/k}\right)^2} \tag{39}$$

Since the solutions of Eq. (1) with potential $V(r)$ are given by Eqs. (22) and (23), the energy spectrum and the eigenfunctions will be

$$E_n = \frac{-1}{4k^2}\left[\frac{\left(n + \frac{(h_L+1)}{2}\right)^2 + k^2(A - C)}{n + \frac{(h_L+1)}{2}}\right]^2 \tag{40}$$

and

$$\psi_{nL}(r) = \left(e^{-r/k}\right)^{\frac{C_L - 1}{2}}\left(1 - e^{-r/k}\right)^{\frac{h_L+1}{2}} {}_2F_1\left(-n, b_L; c_L : e^{-r/k}\right) \tag{41}$$

where the new parameters defined in Eq. (31) have been used. Besides, according with Eqs. (19) and (20)

$$h_L = \sqrt{(2L+1)^2 + 4k^2(B + C)}, \tag{42}$$

$$b_L = \frac{(h_L + 1)(h_L + n) - 2k^2(A + B) - 2L_s}{2n + h_L + 1} \tag{43}$$

and

$$c_L = \frac{-2n(h_L + n) - 2k^2(A + B) - 2L_s}{2n + h_L + 1} \tag{44}$$

Furthermore, the number of states will be determined by Eq. (21) as

$$0 \leq n < k\sqrt{C-A} - \frac{h_L+1}{2} \tag{45}$$

Likewise, in accordance with Eq. (17)

$$V(r_{min}) = \frac{-\left(k^2(A+B)+L_s\right)^2}{4k^2\left(k^2(B+C)+L_s\right)} \tag{46}$$

with

$$r_{min} = kln\left(\frac{k^2(A-B-2C)-L_s}{k^2(A+B)+L_s}\right), \tag{47}$$

such that

$$k^2(A+B) < L_s \leq k^2(B+C) \tag{48}$$

or more explicitly

$$k^2(A+B)+l(l+1) < \frac{(3-D)(4l+D-1)}{4} \leq k^2(B+C)+l(l+1) \tag{49}$$

where all possible values of $l = 0, 1, 2, 3, \ldots l_{max}$ fulfill the above inequality.

With these elements, the KG equation in arbitrary dimensions given in Eq. (30), for $S(r) = V(r) = V(r)$; becomes a Schrödinger-type equation

$$\frac{-d^2\psi_{nL}(r)}{dr^2} + \left\{\Delta EV(r) + \frac{L_s}{r^2}\right\}\psi_{nL}(r) = \tilde{E}_{nL}\psi_{nL}(r) \tag{50}$$

with $\Delta E = 2(E_{nL} + M)$ and $\tilde{E}_{nL} = E_{nL}^2 - M^2$ Hence, within the frame of the standard Green-Aldrich approximation, directly from Eqs. (40) and (41), the energy spectrum and wave function are respectively

$$\tilde{E}_{nL} = \frac{-1}{16k^2}\left[\frac{(2n+h_L+1)^2 + 4k^2\Delta E(A-C)}{2n+h_L+1}\right]^2 \tag{51}$$

$$\psi_{nL}(r) = \left(e^{-r/k}\right)^{\frac{C_L-1}{2}}\left(1-e^{-r/k}\right)^{\frac{h_L+1}{2}} {}_2F_1\left(-n, b_L; c_L : e^{-r/k}\right) \tag{52}$$

where

$$h_L = \sqrt{(2L+1)^2 + 4k^2\Delta E(B+C)}, \tag{53}$$

$$b_L = \frac{(h_L+1)(h_L+n) - 2k^2\Delta E(A+B) - 2L_s}{2n+h_L+1} \tag{54}$$

and

$$c_L = \frac{-2n(h_L + n) - 2k^2 \Delta E(A + B) - 2L_s}{2n + h_L + 1} \tag{55}$$

The usefulness of our alternative approach for the calculation of bound state solutions of the D-dimensional KG equation with exponential-type potentials is exemplified in the next section.

5. Applications

The choice of particular values for the parameters A, B, and C appearing in the multiparameter exponential-type potential of Eq. (39), leads to the solutions of the KG equation in arbitrary dimensions for specific potentials. So, without being exhaustive, at the following we are going to consider only some well-known special cases, it being understood the existence of many others that can be treated in a similar way.

5.1 The Eckart+Hultén potential

If we assume the parameters $A = -(V_2 + V_3); B = 4V_1; C = 0$ and $k = (2a)^{-1}$ the potential $V(r)$ in Eq. (39) will be

$$V(r) = \frac{-(V_2 + V_3)e^{-2\alpha r}}{1 - e^{-2\alpha r}} + \frac{4V_1 e^{-2\alpha r}}{(1 - e^{-2\alpha r})^2} \tag{56}$$

such that the Eckart-type potential which also includes the Hulthén potential is written as

$$V_{EH} = V(r) - V_2 = \frac{-V_2}{1 - e^{-2\alpha r}} - \frac{V_3 e^{-2\alpha r}}{1 - e^{-2\alpha r}} + \frac{4V_1 e^{-2\alpha r}}{(1 - e^{-2\alpha r})^2} + \frac{L_s}{r^2} \tag{57}$$

Hence, from Eq. (51) this potential has an energy spectrum given by

$$\tilde{E}_{nL}^{EH} = \frac{-\alpha^2}{4}\left[\frac{(2n + h_L + 1)^2 + \alpha^{-2}\Delta E(V_2 + V_3)}{2n + h_L + 1}\right]^2 - \Delta E V_2 \tag{58}$$

which agrees with the transcendental Eq. (38) of the reference [14], after considering some algebraic steps on it and the displacement $-V_2$ in the potential $V(r)$. At this point, we want to notice that the term $\Delta E V_2$ in above equation appears from $V(r) = V_{EH} + V_2$ given in Eq. (57). That is,

$$\frac{-d^2\psi_{nL}(r)}{dr^2} + \left[\Delta E(V_{EH} + V_2) + \frac{L_s}{r^2}\right]\psi_{nL}(r) = \tilde{E}_{nL}\psi_{nL}(r) \tag{59}$$

Implies

$$\frac{-d^2\psi_{nL}(r)}{dr^2} + \left[\Delta E V_{EH} + \frac{L_s}{r^2}\right]\psi_{nL}(r) = (\tilde{E}_{nL} - \Delta E V_2)\psi_{nL}(r) \tag{60}$$

Likewise from Eq. (52), the corresponding wave functions are

$$\psi_{nL}(r) = \left(e^{-2\alpha r}\right)^{\frac{c_{EH(L)}-1}{2}}\left(1-e^{-2\alpha r}\right)^{\frac{h_{EH(L)}+1}{2}} {}_2F_1\left(-n, b_{EH(L)}; c_{EH(L)}; e^{-2\alpha r}\right) \tag{61}$$

where

$$h_L = \sqrt{(2L+1)^2 + \alpha^{-2}\Delta E V_1}, \tag{62}$$

$$b_{EH(L)} = \left(h_{EH(L)}+1\right)\left(h_{EH(L)}+n\right)- \tag{63}$$

and

$$c_{EH(L)} = -2n\left(h_{EH(L)}+n\right) - 2 \tag{64}$$

5.2 The standard Hultén potential

In this case, for the potential $V(r)$ given in Eq. (39), one can apply the selection

$$A = -Z\alpha, B = C = 0, k = \alpha^{-1} \tag{65}$$

such that the Hultén potential is

$$V_H = \frac{-Z\alpha e^{-\alpha r}}{1-e^{-\alpha r}} + \frac{L_s}{r^2} \tag{66}$$

Similarly to the above case, from Eq. (51), this potential has an energy spectrum

$$\tilde{E}^H_{nL} = \frac{-\alpha^2}{4}\left[n+L+1-\frac{Z\alpha\Delta E}{n+L+1}\right]^2 \tag{67}$$

that agrees with Eq. (47) of the reference [14] under the identification of their $\nu = (D+2l-1)/2$ with our $L = \nu - 1$ i.e. $\nu = L+1$. Additionally, our $\tilde{E}^H_{nL}$ result coincides with that of Saad [18] when the parameters $V_0 = S_0$ and $q = 1$ are used. Similarly, from Eq. (52), the corresponding wavefunctions will be

$$\psi^H_{nL}(r) = \left(e^{-\alpha r}\right)^{\frac{c_{H(L)}-1}{2}}\left(1-e^{-\alpha r}\right)^{L+1} {}_2F_1\left(-n, b_{H(L)}; c_{H(L)}; e^{-\alpha r}\right) \tag{68}$$

where, according to Eqs. (54) and (55), the b_L and b_L parameters are now

$$b_{H(L)} = \frac{(L+1)(2L+n+1) + Z\alpha^{-1} - L_s}{n+L+1} \tag{69}$$

and

$$c_{H(L)} = \frac{-n(2L+n+1) + Z\alpha^{-1} - L_s}{n+L+1} \tag{70}$$

with $h_{H(L)} = 2L+1$.

5.3 The standard Eckart potential

To obtain this potential model, one selects

$$A = -V_1, B = V_2, C = 0; k = b \tag{71}$$

such that

$$V_E = \frac{-V_1 e^{-r/b}}{1 - e^{-r/b}} + \frac{V_2 e^{-r/b}}{\left(1 - e^{-r/b}\right)^2} + \frac{L_s}{r^2} \tag{72}$$

corresponds to the Eckart potential. So, from Eq. (51), its corresponding energy spectrum results in

$$\tilde{E}_{nL}^{E} = \frac{-1}{16b^2}\left[\frac{\left(2n + h_{E(L)} + 1\right)^2 + 4b^2 V_1 \Delta E}{2n + h_{E(L)} + 1}\right]^2 \tag{73}$$

where

$$h_{E(L)} = \sqrt{(2L+1)^2 + 4b^2 \Delta E V_2}, \tag{74}$$

and the eigenfunctions are

$$\psi_{nL}^{E}(r) = \left(e^{-\alpha r}\right)^{\frac{c_{E(L)}-1}{2}}\left(1 - e^{-\alpha r}\right)^{\frac{h_{E(L)}+1}{2}} {}_2F_1\left(-n, b_{E(L)}; c_{E(L)}; e^{-\alpha r}\right) \tag{75}$$

$$b_{EH(L)} = \frac{\left(h_{E(L)} + 1\right)\left(h_{E(L)} + n\right) - 2b^2 \Delta E (V_2 - V_1) - 2L_s}{2n + h_{E(L)} + 1} \tag{76}$$

and

$$c_{EH(L)} = \frac{-2n\left(h_{E(L)} + n\right) - 2b^2 \Delta E (V_2 - V_1) - 2L_s}{2n + h_{E(L)} + 1} \tag{77}$$

It is worth mentioning that in the particular case $D = 3$, the energy spectrum given in Eq. (73) coincides with the results of Akpan et al. [17] by assuming the standard Green and Aldrich approximation ($C_0 = 0; C_1 = C_2 = 1$).

5.4 The Manning-Rosen potential

Let us consider now the parameters

$$A = -V_0/b^2, B = 0, C = \frac{\alpha(\alpha-1)}{b^2}; k = b \tag{78}$$

from which, one has the Manning-Rosen potential

$$V_{MR} = \frac{1}{b^2}\left(\frac{\alpha(\alpha-1)e^{-2r/b}}{\left(1 - e^{-r/b}\right)^2} - \frac{V_0 e^{-r/b}}{1 - e^{-r/b}}\right) + \frac{L_s}{r^2} \tag{79}$$

with energy spectrum

$$\tilde{E}_{nL}^{MR} = \frac{-1}{16b^2}\left[\frac{\left(2n + h_{MR(L)} + 1\right)^2 + 4\Delta E(V_0 + \alpha(\alpha - 1))}{2n + h_{MR(L)} + 1}\right]^2 \tag{80}$$

where

$$h_{MR(L)} = \sqrt{(2L+1)^2 + 4\Delta E\alpha(\alpha - 1)}, \tag{81}$$

Besides, the eigenfunctions are

$$\psi_{nL}^{MR}(r) = \left(e^{-r/b}\right)^{\frac{c_{MR(L)}-1}{2}}\left(1 - e^{-r/b}\right)^{\frac{h_{MR(L)}+1}{2}} {}_2F_1\left(-n, b_{MR(L)}; c_{MR(L)}; e^{-r/b}\right) \tag{82}$$

$$b_{MR(L)} = \frac{\left(h_{MR(L)} + 1\right)\left(h_{MR(L)} + n\right) + 2\Delta E V_0 - 2L_s}{2n + h_{MR(L)} + 1} \tag{83}$$

and

$$c_{MR(L)} = \frac{-2n\left(h_{MR(L)} + n\right) + 2\Delta E V_0 - 2L_s}{2n + h_{MR(L)} + 1} \tag{84}$$

5.5 The improved Manning-Rosen potential

To get this special case, it becomes necessary the choice $A = -2D_e(e^{\alpha r_e} - 1)$, $B = 0$; $C = D_e(e^{\alpha r_e} - 1)^2$, and $k = \alpha^{-1}$ leading to Improved Manning-Rosen potential

$$V_{IMR}(r) = V(r) + D_e = D_e\left(1 - \frac{e^{\alpha r_e} - 1}{e^{\alpha r} - 1}\right)^2 + \frac{L_s}{r^2} \tag{85}$$

with, according to Eq. (49), energy spectrum given by

$$\tilde{E}_{nL}^{IMR} = -\alpha^2\left[\frac{2n + h_{IMR(L)} + 1}{4} - \frac{\alpha^{-2}D_e\Delta E(e^{2\alpha r_e} - 1)}{2n + h_{IMR(L)} + 1}\right]^2 + D_e\Delta E \tag{86}$$

where

$$h_{IMR(L)} = \sqrt{(2L+1)^2 + 4\alpha^{-2}D_e\Delta E(e^{\alpha r_e} - 1)^2}, \tag{87}$$

At this point, we want to notice that in the case of $D = 3$, the energy spectrum $\tilde{E}_{nL}^{IMR}$ is in agreement with Eq. (31) of Jia et al. [19] besides, it corrects Eqs. (21) and (24) of the reference [20].

On the other hand, from Eq. (52), the eigenfunctions are in this case

$$\psi_{nL}^{IMR}(r) = \left(e^{-\alpha r}\right)^{\frac{c_{IMR(L)}-1}{2}}\left(1 - e^{-\alpha r}\right)^{\frac{h_{IMR(L)}+1}{2}} {}_2F_1\left(-n, b_{IMR(L)}; c_{IMR(L)}; e^{-\alpha r}\right) \tag{88}$$

with

$$b_{IMR(L)} = \frac{\left(h_{IMR(L)} + 1\right)\left(h_{IMR(L)} + n\right) - 4\alpha^{-2}\Delta E(e^{\alpha r_e} - 1) - 2L_s}{2n + h_{IMR(L)} + 1} \tag{89}$$

and

$$c_{IMR(L)} = \frac{-2n\left(h_{IMR(L)} + n\right) + 4\alpha^{-2}\Delta E(e^{\alpha r_e} - 1) - 2L_s}{2n + h_{IMR(L)} + 1} \tag{90}$$

5.6 The Hylleraas potential

Assuming that $A = -V_0(1-a)$, $B = C = 0$ and $k = (2\alpha)^{-1}$ the potential given in Eq. (39) reduces to

$$V(r) = \frac{-V_0(1-a)e^{-2\alpha r}}{1 - e^{-2\alpha r}} \tag{91}$$

for which the Hylleraas potential in D-dimensions is given by

$$V_{Hy} = V(r) + V_0 a = V_0\left(\frac{a - e^{-2\alpha r}}{1 - e^{-2\alpha r}}\right) + \frac{L_s}{r^2} \tag{92}$$

As before, from Eq. (51), the corresponding energy spectrum will be

$$\tilde{E}_{nL}^{Hy} = \frac{-\alpha^2}{4}\left[\frac{\left(2n + h_{Hy(L)} + 1\right)^2 + \alpha^{-2}D_e\Delta E V_0(a-1)}{2n + h_{Hy(L)} + 1}\right]^2 + aV_0\Delta E \tag{93}$$

where

$$h_{Hy(L)} = 2L + 1, \tag{94}$$

in agreement with Hassamaadi et al. [21] when considering their parameters $b = 1$ and $V_1 = V_2 = 0$.

In relation with wavefunctions, from Eq. (52), these are

$$\psi_{nL}^{Hy}(r) = \left(e^{-\alpha r}\right)^{\frac{c_{Hy(L)}-1}{2}}\left(1 - e^{-\alpha r}\right)^{\frac{h_{Hy(L)}+1}{2}} {}_2F_1\left(-n, b_{Hy(L)}; c_{Hy(L)}; e^{-\alpha r}\right) \tag{95}$$

being

$$b_{Hy(L)} = \frac{\left(h_{Hy(L)} + 1\right)\left(h_{Hy(L)} + n\right) - 2\alpha^{-1}\Delta E V_0(a-1) - 2L_s}{2n + h_{Hy(L)} + 1} \tag{96}$$

and

$$c_{Hy(L)} = \frac{-2n\left(h_{Hy(L)} + n\right) - 2\alpha^{-1}\Delta E V_0(a-1) - 2L_s}{2n + h_{Hy(L)} + 1} \tag{97}$$

where, as in all the above cases, the down index indicates the name of potential, i.e. Hy refers to Hylleraas.

5.7 The Deng Fan potential

In this case, the involved parameters are chosen as

$$A = -2bD_e, B = 0, C = D_e b^2 \text{and}\, k = 1/\alpha \tag{98}$$

such that the Deng Fan potential, also called generalized Morse potential will be

$$V_{DF} = \frac{-2bD_e e^{-\alpha r}}{1 - e^{-\alpha r}} + \frac{D_e b^2 e^{-2\alpha r}}{(1 - e^{-\alpha r})^2} + \frac{L_s}{r^2} \tag{99}$$

where D_e is the dissociation energy. The corresponding energy spectrum is obtained from Eq. (51) as

$$\tilde{E}_{nL}^{DF} = \frac{-\alpha^2}{16}\left[\frac{(2n + h_{DF(L)} + 1)^2 - 4\alpha^{-2}D_e\Delta E(2b + b^2)}{2n + h_{DF(L)} + 1}\right]^2 \tag{100}$$

which is in agreement with Oluwadare et al. [22] for the three-dimensional case, when considering the Green and Aldrich approximation [12]. Finally, from Eq. (52), the respective wave functions are

$$\psi_{nL}^{DF}(r) = (e^{-\alpha r})^{\frac{c_{DF(L)}-1}{2}}(1 - e^{-\alpha r})^{\frac{h_{DF(L)}+1}{2}}{}_2F_1(-n, b_{DF(L)}; c_{DF(L)}; e^{-\alpha r}) \tag{101}$$

where

$$h_{DF(L)} = \sqrt{(2L + 1)^2 + 4\alpha^{-2}D_e\Delta E D_e b^2}, \tag{102}$$

$$b_{DF(L)} = \frac{(h_{DF(L)} + 1)(h_{DF(L)} + n) + 4\alpha^{-2}\Delta E b D_e - 2L_s}{2n + h_{DF(L)} + 1} \tag{103}$$

and

$$c_{DF(L)} = \frac{-2n(h_{DF(L)} + n) + 4\alpha^{-2}\Delta E b D_e - 2L_s}{2n + h_{DF(L)} + 1} \tag{104}$$

At this point, it should be noted that the equivalence among the Manning-Rosen potential, the Deng-Fan and Schiöberg models for diatomic molecules have been already shown with detail in references [23, 24]. So, the solutions of the KG equation in arbitrary dimensions derived in this section can also be extended to the Schiöberg potential [25].

Finally, we want to pay attention that in a similar manner to the examples considered in this work, other exponential potentials would be achieved as particular cases from our general proposal of multi-parameter exponential-type potential [26] after a proper selection of the involved parameters.

6. Concluding remarks

Through a direct approach to transform the Schrödinger equation into a hypergeometric differential equation, we have obtained the exact solution of a class of multiparameter exponential-type potentials. Also, we have used the fact that, for equal Lorentz vector and scalar potentials, the Klein-Gordon equation can be written as a Schrödinger-type equation. With these elements, and with a proper redefinition of the involved parameters, we propose an approach to obtain the analytical solutions of the Klein-Gordon equation for exponential-type potentials, in the frame of the Green-Aldrich approximation to the centrifugal term. As a test of the usefulness of the proposed method, by an appropriate selection of parameters, the Klein-Gordon equation has been solved for specific exponential potential models such as Hulthén, Eckart, Manning-Rosen, Improved Manning-Rosen, Hylleraas and generalized Morse or Deng Fan which are derived here as particular cases from the proposal. That is, with this work, we are proposing a unified treatment for solving the Klein-Gordon equation subject to multiparameter exponential-type-potentials, leaving aside the usual methods of solution applied for each one of the aforementioned potentials, for particular parameters, given as examples. So, the displayed method offers an alternative treatment of spinless particles with new exponential-type potentials as well as the possibility to use other schemes of approximations to the centrifugal term.

Acknowledgements

This work was partially supported by the project UAMA-CBI-2232004-009. One of us (JGR) is indebted to the Instituto Politécnico Nacional-Mexico (IPN) for the financial support given through the COFAA-IPN project SIP-20231470. We are grateful to the SNI-Conacyt-México for the stipend received.

Author details

José Juan Peña[1]*, Jesús Morales[1] and Jesús García-Ravelo[2]

1 Metropolitan Autonomous University, Azc. CDMX, Mexico

2 National Polytechnic Institute, ESFM, CDMX, Mexico

*Address all correspondence to: jjpg@azc.uam.mx

References

[1] Nikiforo A, Uvarov V. Special Functions of Mathematical Physics. Bassel: Birkhauser; 1988

[2] Olgar E, Koc R, Tütüncüler H. The exact solution of the s-wave Klein-Gordon equation for the generalized Hulthén potential by the asymptotic iteration method, Phisyca Scripta. 2008; **78**;015011

[3] Ahmadov AI, Nagiyev SM, Qocayeva MV, Uzun K, Tarverdiyeva VA. Bound state solution of the Klein-Fock-Gordon equation with the Hulthén plus a ring-shaped-like potential within SUSY quantum mechanics. International Journal of Modern Physics A. 2018; **33**(33):1850203

[4] Yusufoglu E. The variational iteration method for studying the Klein-Gordon equation. Applied Mathematics Letters. 2008;**21**:669

[5] Chatterjee A. Large-N solution of the Klein-Gordon equation. Journal of Mathematical Physics. 1986;**27**:2331

[6] Sun H. Quantization Rule for Relativistic Klein-Gordon Equation. Bulletin of Korean Chemical Society. 2011;**32**:4233

[7] Bülbül B, Sezer M. A New Approach to Numerical Solution of Nonlinear Klein-Gordon Equation. Mathematical Problems in Engineering. 2013;**869749**:7

[8] Klein O. Quantentheorie und fünfdimensionale Relativitätstheorie. Zeitschrift für Physik. 1926;**37**:895

[9] Gordon W. Der Comptoneffekt nach der Schrödingerschen Theorie. Zeitschrift für Physik. 1926;**40**:117

[10] Okorie US, Ikot AN, Onate CA, Onyeaju MC, Rampho GJ. Bound and scattering states solutions of the Klein-Gordon equation with the attractive radial potential in higher dimensions. Modern Physics Letters A. 2021;**36**(32): 2150230

[11] Peña J, Morales J, García-Ravelo J. Bound state solutions of Dirac equation with radial exponential-type potentials. Journal of Mathematical Physics. 2017; **48**:043501

[12] Nath D, Roy AK. Analytical solution of D dimensional Schrödinger equation for Eckart potential with a new improved approximation in centrifugal term. Chemical Physics Letters. 2021; **780**:138909

[13] Dhahbi A, Landolsi AA. The Klein-Gordon equation with equal scalar and vector Bargmann potentials in D dimensions. Results in Physics. 2022;**33**: 105143

[14] Ikhdair SM. Bound state energies and wave functions of spherical quantum dots in presence of a confining potential model. Journal of Quantum Information Science. 2011;**1**:73

[15] Peña JJ, García-Martínez J, García-Ravelo J, Morales J. Bound state solutions of D-dimensional schrödinger equation with exponential-type potentials. International Journal of Quantum Chemistry. 2015;**115**:158

[16] Jia CS, Diao YF, Yi LZ, Chen T. Arbitrary l-wave Solutions of the Schrödinger Equation with The Hultén Potential Model. International Journal of Modern Physics A. 2009; **24**:4519

[17] Akpan IO, Antia AD, Icot AN. Bound-State Solutions of the

Klein-Gordon Equation with q-Deformed Equal Scalar and Vector Eckart Potential Using a Newly Improved Approximation Scheme. ISRN High Energy Physics. 2012. ID 798209

[18] Saad N. The Klein-Gordon equation with a generalized Hulthén potential in D-dimensions. Physica Scripta. 2007; **76**:623

[19] Jia CS, Chen T, He S. Bound state solutions of the Klein-Gordon equation with the improved expression of the Manning-Rosen potential energy model. Physics Letters A. 2013;**377**:682

[20] Chen XY, Chen T, Jia CS. Solutions of the Klein-Gordon equation with the improved Manning-Rosen potential energy model in D dimensions. The European Physical Journal Plus. 2014; **129**:75

[21] Hassanabadi S, Maghsoodi E, Oudi R, Sarrinkamar S, Rahimov H. Exact solution Dirac equation for an energy-dependent potential. The European Physical Journal Plus. 2012; **127**:120

[22] Oluwadare OJ, Oyewumi KJ, Akoshile CO, Babalola OA. Approximate analytical solutions of the relativistic equations with the Deng-Fan molecular potential including a Pekeris-type approximation to the (pseudo or) centrifugal term. Physica Scripta. 2012; **86**:035002

[23] Wang PQ, Zhang LH, Jia CS, Liu JY. Equivalence of the three empirical potential energy models for diatomic molecules. Journal of Molecular Spectroscopy. 2012;**274**:5

[24] Peña J, Ovando G, Morales J. On the equivalence of radial potential models for diatomic molecules. Theoretical Chemistry Accounts. 2016;**135**:62

[25] Omugbe E, Osafile OE, Okon IB, Enaibe EA, Onyeaju MC. Bound state solutions, Fisher information measures, expectation values, and transmission coefficient of the Varshni potential. Molecular Physics, 2021;**119**:e1909163

[26] Onate CA, Onyeaju MC, Okon IB, Adeoti A. Molecular energies of a modified and deformed exponential-type potential model. Chemical Physics Impact. 2021;**3**:100045

Chapter 6

From a 4-Rank Totally Antisymmetric Field Strength to Two Dual Electromagnetic Fields in Four Time and Four Space Dimensions

Juan Antonio Nieto

Abstract

A 4-rank "electromagnetic" gauge field strength in four-time and four-space dimensions ($(4+4)$-dimensions) is considered. We show that by necessitating such a four-rank gauge field we satisfy the Grassmann–Plücker relations which allows to choose an ansatz (for a basic basis) such that it is broken into two dual electromagnetic fields – one in a $(1+3)$-world and the other in a $(3+1)$-world. An interesting aspect of this mechanism is that the electromagnetic field $(1+3)$-world turns out to be dual to the electromagnetic field in the $(3+1)$-world.

Keywords: 4-rank field strength, (4 + 4)-dimensions, electromagnetic field, Maxwell equations and higher dimensional theory, antisymmetric field strength

1. Introduction

It is known that the Grassmann Plücker relations [1–6] of totally antisymmetric forms determine the Plücker coordinates which mean that such a totally antisymmetric form is decomposable (see Ref. [3] and references therein). Physically, the Plücker embedding can be found in a Grassmannian sigma model in $SU(2)$ Yang Mills model [7] and in spherically symmetric instantons of the scale invariant $SU(2)$ gauged Grassmannian model in $d = 4$ [8]. When this developments are applied to 2-rank antisymmetric gauge field in four-dimensions, it is found that the corresponding electromagnetic field strength can be written in terms of the true degrees of freedom [9].

In this work, we make a number of remarks on Plücker coordinates associated with 4-rank totally antisymmetric gauge fields strength $F^{\hat{\mu}\hat{\nu}\hat{\alpha}\hat{\beta}}$ (differential 4-form) in $(4+4)$-dimensions. When such a gauge field satisfies the Grassmann–Plücker relations can be, of course, decomposable in terms of more elementary quantities. Surprisingly, for a particular case of ansatz for such elementary basic quantities, the field equations for $F^{\hat{\mu}\hat{\nu}\hat{\alpha}\hat{\beta}}$ lead to both the electromagnetic field equation for $F^{\mu\nu}$ in $(1+3)$-dimensions and the electromagnetic field equation for G^{ij} in $(3+1)$-dimensions. An interesting aspect of this mechanism is that the source of $F^{\mu\nu}$ is determined in part by G^{ij} and dually the source of G^{ij} is determined in part by $F^{\mu\nu}$.

2. Field equations for a 4-rank field strength

Let us start considering a totally antisymmetric gauge field strength

$$F^{\hat{\mu}\hat{\nu}\hat{\alpha}\hat{\beta}} = F^{\hat{\mu}\hat{\nu}\hat{\alpha}\hat{\beta}}\left(x^{\sigma}, y^{i}\right)$$

in $(4+4)$-dimensions, which we shall assume is a function of $(1+3)$-coordinates x^{σ} and $(3+1)$-coordinates y^{i}. Suppose that $F^{\hat{\mu}\hat{\nu}\hat{\alpha}\hat{\beta}}$ satisfies the Maxwell type equations:

$$\partial_{\hat{\beta}} F^{\hat{\mu}\hat{\nu}\hat{\alpha}\hat{\beta}} = 0 \tag{1}$$

and

$$\partial_{\hat{\beta}} {}^{*}F^{\hat{\mu}\hat{\nu}\hat{\alpha}\hat{\beta}} = 0, \tag{2}$$

where ${}^{*}F^{\hat{\mu}\hat{\nu}\hat{\alpha}\hat{\beta}}$ is the dual gauge field defined as

$${}^{*}F^{\hat{\mu}\hat{\nu}\hat{\alpha}\hat{\beta}} = \frac{1}{4!}\epsilon^{\hat{\mu}\hat{\nu}\hat{\alpha}\hat{\beta}\hat{\sigma}\hat{\rho}\hat{\gamma}\hat{\eta}} F_{\hat{\sigma}\hat{\rho}\hat{\gamma}\hat{\eta}}. \tag{3}$$

In general, we raise and lower indices with a flat Minkowski metric $\eta_{\hat{\mu}\hat{\nu}}$, which in $(4+4)$-dimensions takes the form

$$\eta_{\hat{\mu}\hat{\nu}} = diag(-1,1,1,1,-1,-1,-1,1). \tag{4}$$

The ϵ-symbol is a totally antisymmetric symbol (Levi–Civita symbol) defined as

$$\epsilon^{\hat{\mu}\hat{\nu}\hat{\alpha}\hat{\beta}\hat{\sigma}\hat{\rho}\hat{\gamma}\hat{\eta}} \in \{-1,0,1\}. \tag{5}$$

In fact, the ϵ-symbol has values $+1$ or -1 depending on even or odd permutations of $\epsilon^{12\ldots8}$, respectively, otherwise the ϵ-symbol is zero. It is verified that the relation

$$\epsilon^{\hat{\mu}_1\ldots\hat{\mu}_8}\epsilon_{\hat{\nu}_1\ldots\hat{\nu}_8} = -\delta^{\hat{\mu}_1\ldots\hat{\mu}_8}_{\hat{\nu}_1\ldots\hat{\nu}_8}, \tag{6}$$

where $\delta^{\hat{\mu}_1\ldots\hat{\mu}_8}_{\hat{\nu}_1\ldots\hat{\nu}_8}$ is a generalized Kronecker delta [10].

Let us assume that the Grassmann–Plücker relations (see Ref. [3] and references therein) hold for $F^{\hat{\mu}\hat{\nu}\hat{\alpha}\hat{\beta}}$, namely

$$F^{\hat{\mu}\hat{\nu}\hat{\alpha}\left[\hat{\beta}\right.}F^{\left.\hat{\sigma}\hat{\rho}\hat{\gamma}\hat{\eta}\right]} = 0. \tag{7}$$

Here, the bracket $\left[\hat{\beta}\hat{\sigma}\hat{\rho}\hat{\gamma}\hat{\eta}\ldots\right]$ means totally antisymmetric. This implies that $F^{\hat{\mu}\hat{\nu}\hat{\alpha}\hat{\beta}}$ is decomposable. In other words, this means that $F^{\hat{\mu}\hat{\nu}\hat{\alpha}\hat{\beta}}$ can be written as

$$F^{\hat{\mu}\hat{\nu}\hat{\alpha}\hat{\beta}} = \frac{1}{4!}\varepsilon^{\hat{A}\hat{B}\hat{C}\hat{D}} v^{\hat{\mu}}_{\hat{A}} v^{\hat{\nu}}_{\hat{B}} v^{\hat{\alpha}}_{\hat{C}} v^{\hat{\beta}}_{\hat{D}}. \tag{8}$$

Here, the elementary quantities $v^{\hat{\mu}}_{\hat{A}} = v^{\hat{\mu}}_{\hat{A}}\left(x^{\sigma}, y^{i}\right)$ can be considered as basic basis elements and $\varepsilon^{\hat{A}\hat{B}\hat{C}\hat{D}}$ is an 'internal' four-dimensional ϵ-symbol.

Moreover, from (2) one learns that

$$F_{\hat{\mu}\hat{\nu}\hat{\alpha}\hat{\beta}} = \partial_{\hat{\mu}} A_{\hat{\nu}\hat{\alpha}\hat{\beta}]}. \tag{9}$$

So one sees that $A_{\hat{\nu}\hat{\alpha}\hat{\beta}}$ is a totally antisymmetric gauge field which under the transformation

$$A_{\hat{\nu}\hat{\alpha}\hat{\beta}} \to A_{\hat{\nu}\hat{\alpha}\hat{\beta}} + \partial_{\hat{\nu}} \Omega_{\hat{\alpha}\hat{\beta}]}, \tag{10}$$

the 4-rank tensor $F_{\hat{\mu}\hat{\nu}\hat{\alpha}\hat{\beta}}$ becomes invariant.

3. Kaluza-Klein type ansatz

In general, one finds that $F^{\hat{\mu}\hat{\nu}\hat{\alpha}\hat{\beta}}$ can be written as

$$\begin{aligned} F^{\hat{\mu}\hat{\nu}\hat{\alpha}\hat{\beta}} &= F^{\hat{\mu}\hat{\nu}} G^{\hat{\alpha}\hat{\beta}} - F^{\hat{\mu}\hat{\alpha}} G^{\hat{\nu}\hat{\beta}} + F^{\hat{\mu}\hat{\beta}} G^{\hat{\nu}\hat{\alpha}} \\ &+ G^{\hat{\mu}\hat{\nu}} F^{\hat{\alpha}\hat{\beta}} - G^{\hat{\mu}\hat{\alpha}} F^{\hat{\nu}\hat{\beta}} + G^{\hat{\mu}\hat{\beta}} F^{\hat{\nu}\hat{\alpha}}, \end{aligned} \tag{11}$$

where

$$F^{\hat{\mu}\hat{\nu}} = \frac{1}{4!}\epsilon^{ab} v^{\hat{\mu}}_{a} v^{\hat{\nu}}_{b} \tag{12}$$

and

$$G^{\hat{\mu}\hat{\nu}} = \frac{1}{4!}\epsilon^{AB} v^{\hat{\mu}}_{A} v^{\hat{\nu}}_{B}. \tag{13}$$

Our next step is to consider some particular cases. In principle one may assume a kind of Kaluza–Klein ansatz for $v^{\hat{\mu}}_{\hat{A}}$, namely

$$v^{\hat{\mu}}_{\hat{A}} = \begin{pmatrix} v^{\mu}_{a} & v^{\mu}_{A} \\ 0 & v^{i}_{A} \end{pmatrix}. \tag{14}$$

However, in this case, one looses the symmetry between the $(1+3)$-world and the $(3+1)$-world. So one shall assume that

$$v^{\hat{\mu}}_{\hat{A}} = \begin{pmatrix} v^{\mu}_{a} & 0 \\ 0 & v^{i}_{A} \end{pmatrix}. \tag{15}$$

In this case, one gets

$$F^{\mu i} = 0 \tag{16}$$

and

$$F^{ij} = 0. \tag{17}$$

And one also obtains

$$G^{\mu i} = 0 \tag{18}$$

and

$$G^{\mu\nu} = 0. \tag{19}$$

Hence, using Eq. (11) one learns that the only nonvanishing component of $F^{\hat{\mu}\hat{\nu}\hat{\alpha}\hat{\beta}}$ is

$$F^{\mu\nu ij} = F^{\mu\nu}\left(x^{\sigma}, y^{k}\right) G^{ij}\left(x^{\lambda}, y^{k}\right), \tag{20}$$

where the "electromagnetic" fields $F_{\mu\nu}$ and G_{ij} corresponds to the $(1+3)$-world and $(3+1)$-world, respectively.

4. Inhomogeneous Maxwell field equations

From (1) one knows that

$$\partial_{\beta} F^{\hat{\mu}\hat{\nu}\hat{\alpha}\beta} + \partial_{j} F^{\hat{\mu}\hat{\nu}\hat{\alpha}j} = 0. \tag{21}$$

Thus, the relevant equations that can be obtained from Eq. (21) are

$$\partial_{\nu} F^{\mu\nu ij} = 0 \tag{22}$$

and

$$\partial_{j} F^{\mu\nu ij} = 0. \tag{23}$$

Hence, using Eq. (20) one learns that

$$(\partial_{\nu} F^{\mu\nu}) G^{ij} + \left(\partial_{\nu} G^{ij}\right) F^{\mu\nu} = 0 \tag{24}$$

and

$$\left(\partial_{j} G^{ij}\right) F^{\mu\nu} + \left(\partial_{j} F^{\mu\nu}\right) G^{ij} = 0. \tag{25}$$

These equations can also be written as

$$\partial_{\nu} F^{\mu\nu} = J^{\mu} \tag{26}$$

and

$$\partial_{j} G^{ij} = J^{i}. \tag{27}$$

Here,

$$J^{\mu} = -F^{\mu\nu}\partial_{\nu}\ln\psi \tag{28}$$

and

$$J^{i} = -G^{ij}\partial_{j}\ln\xi, \tag{29}$$

with

$$\psi = \left(\frac{1}{2}G^{ij}G_{ij}\right)^{1/2} \tag{30}$$

and

$$\xi = \left(\frac{1}{2}F^{\mu\nu}F_{\mu\nu}\right)^{1/2}. \tag{31}$$

Another interesting way to write (26) and (27) is

$$\partial_{\nu}(\psi F^{\mu\nu}) = 0 \tag{32}$$

and

$$\partial_{j}(\xi G^{ij}) = 0, \tag{33}$$

respectively.

It is evident that (26) (and (27)) or (32) (and (33)) are inhomogeneous Maxwell field type equations. Consequently, $F^{\mu\nu}$ can be identified with the electromagnetic field strength in the $(1+3)$-world and G^{ij} with the dual-mirror electromagnetic field strength in the $(3+1)$-world. However, (26) or (32) establishes something else that the source of the electromagnetic field ψ arises in part from the $(3+1)$-world *via* G^{ij}, and the eqs. (27) and (33) indicate that the source of the mirror electromagnetic field emerges from the $(1+3)$-world *via* $F^{\mu\nu}$. This process shows a duality between the $(1+3)$-world and the $(3+1)$-world. An important point is that both electromagnetic fields $F^{\mu\nu}$ and G^{ij} are part according to (18) of the 4-rank gauge field strength $F_{\hat{\mu}\hat{\nu}\hat{\alpha}\hat{\beta}}$ which "lives" in a $(4+4)$-world.

5. Dual Maxwell field equations

Moreover, from Eq. (20) one can also show that the only nonvanishing components of the dual gauge field strength ${}^{*}F^{\hat{\mu}\hat{\nu}\hat{\alpha}\hat{\beta}}$ are ${}^{*}F^{\mu\nu ij}$ which can be written as

$${}^{*}F^{\mu\nu ij} = {}^{*}F^{\mu\nu}\left(x^{\sigma}, y^{k}\right){}^{*}G^{ij}\left(x^{\lambda}, y^{k}\right), \tag{34}$$

where

$${}^{*}F^{\mu\nu} = \frac{1}{2!}\epsilon^{\mu\nu\alpha\beta}F_{\alpha\beta} \tag{35}$$

and

$$^*G^{ij} = \frac{1}{2!}\epsilon^{ijkl}G_{kl}. \tag{36}$$

From the field eq. (2), one obtains

$$\partial_\nu{}^*F^{\mu\nu} = -\partial_\nu(\ln\psi^*)^*F^{\mu\nu} \tag{37}$$

and

$$\partial_j{}^*G^{ij} = -\partial_j(\ln\xi^*)^*G^{ij}. \tag{38}$$

Here,

$$\psi^* = \left(\frac{1}{2}^* G^{ij*}G_{ij}\right)^{1/2} \tag{39}$$

and

$$\xi^* = \left(\frac{1}{2}^* F^{\mu\nu*}F_{\mu\nu}\right)^{1/2}. \tag{40}$$

One finds that (37) and (38) can also be written as

$$\partial_\nu(^*\psi^*F^{\mu\nu}) = 0 \tag{41}$$

and

$$\partial_j(^*\xi^*G^{ij}) = 0, \tag{42}$$

respectively.

6. Final remarks

Let us make some final remarks. Usually, for describing different phenomena in our universe, one time and three-space dimensions ($(1+3)$-dimensions) are the chosen number of real dimensions. But the question emerges, why $(1+3)$-dimensions? why not $(3+1)$-dimensions? or why not $(4+4)$-dimensions? Unfortunately (or fortunately) until now these questions are an open theoretical problem; as far as one knows, nobody knows the answer. It turns out that looking for a possible solution one stumbling with the discovery that there is a triality relation between the signatures $(1+9)$, $(5+5)$ and $(9+1)$ [11, 12]. This means that by triality the $(5+5)$-dimensional world can always be related to the other basic signatures $(1+9)$ and $(9+1)$. It turns out that $(5+5)$-world is a common signature to both type *IIA* strings and type *IIB* strings. From this perspective, one may say that the $(4+4)$-world can be considered as the transverse world of the $(5+5)$-world (see Refs [11, 12] and references therein). Moreover, it turns out that in $(4+4)$-dimensions, there are a number of remarkable mathematical and physical results that are worth mentioning.

Mathematically, it has been suggested that the mathematical structures of oriented matroid theory [13] (see Refs. [14–22] and references therein) and surreal number theory [23, 24] (see also Refs [25, 26] and references therein) may provide interesting routes for a connection with the $(4+4)$-world. Physically, the Dirac equation in $(4+4)$-dimensions is consistent with Majorana–Weyl spinors which give exactly the same number of components as the complex spinor of $\frac{1}{2}$-spin particles such as the electron or the quarks (see Refs. [12, 27]). Second, the most general Kruskal–Szekeres transformation of a black-hole coordinates in $(1+3)$-dimensions leads to eight-regions (instead of the usual four-regions), which can be better described in $(4+4)$-dimensions [28]. Third, it also has been shown [29] that duality

$$\sigma^2 \leftrightarrow \frac{1}{\sigma^2}, \tag{43}$$

of a Gaussian distribution in terms of the standard deviation σ of 4-space coordinates associated with the de Sitter space (anti-de Sitter) and the vacuum zero-point energy yields to a Gaussian of 4-time coordinates of the same vacuum scenario. Moreover, loop quantum gravity in $(4+4)$-dimensions [30, 31] admits a self-duality curvature structure analogue to the traditional $(1+3)$-dimensions.

In the above sense, the contribution of this work adds to the fact that "electromagnetic" field in a $(4+4)$-world described by a 4-rank totally antisymmetric field strength $F^{\hat{\mu}\hat{\nu}\hat{\alpha}\hat{\beta}}$ can be broken into two electromagnetic field strengths; the field strength $F^{\mu\nu}$ associated with the $(1+3)$-world and the field strength G^{ij} associated with the $(3+1)$-world. An interesting aspect of this result is that there is a hidden duality symmetry feature of $F^{\mu\nu}$ and G^{ij} in the sense that G^{ij} contribute to the source of $F^{\mu\nu}$ and vice versa.

Finally, it is worth mentioning that 4-rank totally antisymmetric field strength $F^{\hat{\mu}\hat{\nu}\hat{\alpha}\hat{\beta}}$ in $(1+10)$-dimensions are a key mathematical notion in 3-brane theory which, it is known, is an important part in the so-called M-theory (see Ref. [15] and references therein). In fact, in Ref. [9] it is shown how totally antisymmetric fields can be related to p-brane. For the case of the field strength $F^{\hat{\mu}\hat{\nu}\hat{\alpha}\hat{\beta}}$, one uses (6) and writes

$$F^{\hat{\nu}\hat{\alpha}\hat{\beta}\hat{A}} = \frac{1}{4!}\varepsilon^{\hat{B}\hat{C}\hat{D}\hat{A}} v^{\hat{\nu}}_{\hat{B}} v^{\hat{\alpha}}_{\hat{C}} v^{\hat{\beta}}_{\hat{D}} \tag{44}$$

and assume that $F^{\hat{\mu}\hat{\nu}\hat{\alpha}\hat{\beta}} = F^{\hat{\mu}\hat{\nu}\hat{\alpha}\hat{\beta}}\left(x^{\sigma}, y^{i}, \xi^{\hat{A}}\right)$. Moreover, instead of (1) one considers the field equation:

$$\partial_{\hat{A}} F^{\hat{\mu}\hat{\nu}\hat{\alpha}\hat{A}} = 0, \tag{45}$$

where $\partial_{\hat{A}} = \frac{\partial}{\xi^{\hat{A}}}$. This expression implies that due to (44) one can write

$$v^{\hat{\mu}}_{\hat{B}} = \partial_{\hat{B}} X^{\hat{\mu}}. \tag{46}$$

Thus, substituting this result into (8) leads to

$$F^{\hat{\mu}\hat{\nu}\hat{\alpha}\hat{\beta}} = \frac{1}{4!}\varepsilon^{\hat{A}\hat{B}\hat{C}\hat{D}} \partial_{\hat{A}} X^{\hat{\mu}} \partial_{\hat{B}} X^{\hat{\nu}} \partial_{\hat{C}} X^{\hat{\alpha}} \partial_{\hat{D}} X^{\hat{\beta}}. \tag{47}$$

In turn, this expression can be used to write the Schild type action for the 3-brane in target $(4+4)$-dimensions, namely

$$S = \int \left(F^{\hat{\mu}\hat{\nu}\hat{\alpha}\hat{\beta}} F_{\hat{\mu}\hat{\nu}\hat{\alpha}\hat{\beta}} \right)^{1/2} d^4 \xi. \tag{48}$$

So it may be interesting for further work to continue relating our present approach with p-brane theory and M-theory.

Acknowledgements

I would like to thank the Mathematical, Computational & Modeling Sciences Center of the Arizona State University where part of this work was developed.

Classification

Pacs numbers: 04.20.Jb, 04.50.-h, 04.60.-m, 11.15.-q.

Author details

Juan Antonio Nieto[1,2]

1 Facultad de Ciencias Fsico-Matemáticas Universidad Autónoma de Sinaloa Culiacán, Sinaloa, México

2 Facultad de Ciencias de la Tierra y el Espacio Universidad Autónoma de Sinaloa Culiacán, Sinaloa, México

*Address all correspondence to: niet@uas.edu.mx; janieto1@asu.edu

References

[1] Plücker J. On a new geometry of space. Proceedings. Royal Society of London. 1865;**14**:53

[2] Hodge W, Pedoe D. Methods of Algebraic Geometry, 2. American Branch, New York, Cambridge University Press; 1952

[3] Bokowski JLJ, Sturmfels B. Computational synthetic geometry. In: Dold A, Eckmann B, editors. Lecture Notes in Mathematics. New York: Springer-Verlag; 1980

[4] Kolhatkar R. Grassmann Varieties, Thesis. Montreal, Quebec, Canada: Department of Mathematics and Statistics, McGill University; 2004

[5] Sturmfels B. Algorithms in Invariant Theory, Texts and Monographs in Symbolic Computation. Vienna: Springer Verlag; 1993

[6] Baralic D. How to understand Grassmannians? Teaching Mathamatics. 2011;**XIV**:147

[7] Marsh D. The Grassmannian sigma model in SU (2) Yang-Mills theory. Journal of Physics A. 2007;**40**:9919. hep-th/0702134

[8] Chakrabarti A, Tchrakian DH. Spherically symmetric instantons of the scale invariant SU (2) gauged Grassmannian model in d = 4. Physics Letters B. 1996;**376**:59. e-Print: hep-th/9601157

[9] Nieto JA, Nieto-Marín PA, León EA, Garca-Manzanárez E. Remarks on Plücker embedding and totally antisymmetric gauge fields. Modern Physics Letters A. 2020;**35**(22):2050184

[10] Misner CW, Thorne KS, Wheeler JA. Gravitation. San Francisco: W. H. Freeman and Company; 1971

[11] De Andrade MA, Rojas M, Toppan F. Triality of Majorana-Weyl space-times with different signatures. International Journal of Modern Physics A: Particles and Fields; Gravitation; Cosmology; Nuclear Physics. 2001;**16**:4453. hep-th/0005035

[12] Rojas M, De Andrade MA, Colatto LP, Matheus-Valle JL, De Assis LPG and Helayel-Neto JA. Mass Generation and Related Issues from Exotic Higher Dimensions. hep-th/1111.2261

[13] Björner A, Las Vergnas M, Sturmfels B, White N, Ziegler GM. Oriented Matroids. Cambridge: Cambridge University Press; 1993

[14] Nieto JA. Advances in Theoretical and Mathematical Physics. 2004;**8**:177. arXiv: hep-th/0310071

[15] Nieto JA. Advances in Theoretical and Mathematical Physics. 2006;**10**:747. arXiv: hep-th/0506106

[16] Nieto JA. Searching for a connection between matroid theory and string theory. Journal of Mathematical Physics. 2004;**45**:285. arXiv: hep-th/0212100

[17] Nieto JA. Phirotopes, super p-branes and qubit theory. Nuclear Physics B. 2014;**883**:350. arXiv:1402.6998 [hep-th]

[18] Nieto JA, Marin MC. Matroid theory and Chern-Simons. Journal of Mathematical Physics. 2000;**41**:7997. hep-th/0005117

[19] Nieto JA. Qubits and oriented matroids in four time and four space dimensions. Physics Letters B. 2013;**718**: 1543. e-Print: arXiv:1210.0928 [hep-th]

[20] Nieto JA. Qubits and chirotopes. Physics Letters B. 2010;**692**:43. e-Print: arXiv:1004.5372 [hep-th]

[21] Nieto JA. Duality, Matroids, Qubits, Twistors, and Surreal Numbers. Frontiers in Physics. 2018;**6**:106

[22] Nieto JA, Leon EA. 2D Gravity with Torsion, Oriented Matroids and 2+2 Dimensions. Brazilian Journal of Physics. 2010;**40**:383. e-Print: 0905.3543 [hep-th]

[23] Conway JH. 24-28 Oval Road. London NW1. England. On Number and Games, London Mathematical Society Monographs. London, England: Academic Press; 1976

[24] Gonshor H. An introduction to the theory of surreal numbers. In: London Mathematical Society Lectures Notes Series. Vol. 110. Cambridge, UK: Cambridge University Press; 1986

[25] Avalos-Ramos C, Felix-Algandar JA, Nieto JA. Dyadic Rationals and Surreal Number Theory. IOSR Journal of Mathematics. 2020;**16**:35

[26] Nieto JA. Some Mathematical and Physical Remarks on Surreal Numbers. Journal of Modern Physics. 2016;**7**:2164

[27] Nieto JA. Dirac equation in four time and four space dimensions. International Journal of Geometric Methods in Modern Physics. 2016;**14**(01):1750014

[28] Nieto JA, Madriz E. Aspects of (4 + 4)-Kaluza-Klein type theory. Physica Scripta. 2019;**94**:115303

[29] Medina M, Nieto JA, Nieto-Marín PA. Cosmological Duality in Four Time and Four Space Dimensions. Journal of Modern Physics. 2021;**12**:1027

[30] Nieto JA. Superfield description of a self-dual supergravity in the context of MacDowell-Mansouri theory. Classical and Quantum Gravity. 2006;**23**:4387. e-Print: hep-th/0509169

[31] Nieto JA. Towards an Ashtekar formalism in eight dimensions. Classical and Quantum Gravity. 2005;**22**:947. e-Print: hep-th/0410260

Chapter 7

Qubit Lattice Algorithms Based on the Schrödinger-Dirac Representation of Maxwell Equations and Their Extensions

George Vahala, Min Soe, Efstratios Koukoutsis, Kyriakos Hizanidis, Linda Vahala and Abhay K. Ram

Abstract

It is well known that Maxwell equations can be expressed in a unitary Schrodinger-Dirac representation for homogeneous media. However, difficulties arise when considering inhomogeneous media. A Dyson map points to a unitary field qubit basis, but the standard qubit lattice algorithm of interleaved unitary collision-stream operators must be augmented by some sparse non-unitary potential operators that recover the derivatives on the refractive indices. The effect of the steepness of these derivatives on two-dimensional scattering is examined with simulations showing quite complex wavefronts emitted due to transmissions/reflections within the dielectric objects. Maxwell equations are extended to handle dissipation using Kraus operators. Then, our theoretical algorithms are extended to these open quantum systems. A quantum circuit diagram is presented as well as estimates on the required number of quantum gates for implementation on a quantum computer.

Keywords: Schrodinger-Dirac, qubit lattice algorithm, Dyson map, 2D electromagnetic scattering, dissipative systems, Kraus operators, dilation

1. Introduction

Qubit lattice algorithms (QLA) were first being developed in the late 1990s to solve the Schrodinger equation [1–3] using unitary collision and streaming operators acting on some qubit basis. QLA recovers the Schrodinger equation in the continuum limit to second order in the spatial lattice grid spacing. Because the lattice node qubits are entangled by the unitary collision operator (much like in the formation of Bell states), QLA is encodable onto a quantum computer with an expected exponential speed-up over a classical algorithm run on a supercomputer. Moreover, since QLA is extremely parallelizable on a classical supercomputer, it provides an alternate algorithm for solving difficult problems in computational classical physics.

We then applied these QLA ideas to the study of the nonlinear Schrodinger equation (NLS) [4], by incorporating the cubic nonlinearity in the wave function, $|\psi|^2\psi$, as an external potential operator following the unitary collide-stream operator sequence on the qubits. While the inclusion of such nonlinear terms poses no problem for a hybrid classical-quantum computer, it remains a very important and difficult research topic for their implementation on a quantum computer. The accuracy of the QLA for NLS was tested for soliton-soliton collisions in long-term integration and compared to exact analytic solutions, and while the QLA is second order, it seemed to behave like a symplectic integrator. The QLA was then extended to the totally integrable vector Manakov solitons [5] to handle inelastic soliton scattering. The Manakov solitons are solutions to a coupled set of NLS equations.

Following these successful benchmarking simulations, we moved into QLA for two (2D) and three (3D) dimensional NLS equations—where now there are no exact solutions to these nonlinear equations. In the field of condensed matter, these higher dimensions NLS equations are known as the Gross-Pitaevskii equations and give the mean field representation of the ground state wave function ψ of a zero-temperature Bose-Einstein condensate (BEC). For scalar quantum turbulence in 3D, we [6] observed a triple energy cascade on a 5732^3 grid, with the low-k ("classical") regime exhibiting a Kolmogorov $k^{-5/3}$ cascade in the *compressible* kinetic energy while the incompressible kinetic energy exhibited a long k-range of k^{-3} spectrum. Similar results were found for both 2D and 3D scalar quantum [7–9], while results for spinor BECs can be found in Refs. [10–12]. A somewhat related, but significantly different, approach is that of the quantum lattice Boltzmann method [13, 14].

Here we will discuss a QLA for the solution of Maxwell equations in a tensor dielectric medium [15–18] and present some simulation results of the scattering of a 1D electromagnetic pulse off 2D localized dielectric objects. This can be viewed as a precursor to examining the scattering of electromagnetic pulses off plasma blobs in the exterior region of a tokamak.

There has been much interest in rewriting the Maxwell equations in operator form and exploit their similarity to the Schrodinger-Dirac equation from the early 1930s (e.g., see the references in [19]). For homogeneous media, the qubit representation of the electric and magnetic fields, **E, H**, leads to a Dirac equation in a fully unitary representation. However, when the media becomes inhomogeneous, a Dyson map [20] is required to yield a unitary Schrodinger-Dirac equation for the evolution of the electromagnetic qubit field representation. In particular, one can use the fields $(n_xE_x, n_yE_y, n_zE_z, B_x, B_y, B_z)$, where n_i is the refractive index in the ith-direction.

A QLA is developed for this representation of the Maxwell equations in Section 3. This particular algorithm is a generalization of that used for the NLS equations. The initial value problem is then solved for the case of an electromagnetic pulse propagating in the x-direction and scattering from different 2D localized dielectric objects with refractive index $n(x,y)$ in Section 4. In particular, we have examined both polarizations of the pulse and $\nabla \cdot \mathbf{B} = 0$. In Section 5, we consider the case in which the medium is dissipative. This brings in the field of open quantum systems and interactions with an environment. For illustration, we consider a simplified cold electron-ion dissipative fluid model in an electromagnetic field. Kraus operators are determined by a multidimensional analog of the quantum amplitude damping channel. Some estimates on the quantum gates required are given as well as a quantum circuit diagram illustrating the implementation of Kraus operators on the open Schrodinger equation. We summarize our results in Section 6.

Finally, in this introduction, we quickly review the entanglement of qubits and in particular for the 2-qubit Bell state [21].

$$B_+ = \frac{1}{\sqrt{2}}(|00\rangle + |11\rangle). \quad (1)$$

The most general 1-qubit states are $\{a_0|0\rangle + a_1|1\rangle\}$, and $\{b_0|0\rangle + b_1|1\rangle\}$ with normalization $|a_0|^2 + |a_1|^2 = 1 = |b_0|^2 + |b_1|^2$. The tensor product of these two 1-qubit yields a space of the form

$$a_0b_0|00\rangle + a_0b_1|01\rangle + a_1b_0|10\rangle + a_1b_1|11\rangle. \quad (2)$$

However, the Bell state, Eq. (1) is not part of this tensor product space: to remove the $|01\rangle$ state from Eq. (2) either $a_0 = 0$ or $b_1 = 0$. This in turn would remove either the $|00\rangle$ or the $|11\rangle$ states, respectively. Now consider the unitary collision operator

$$C = [\cos\theta \sin\theta - \sin\theta \cos\theta] \quad (3)$$

acting on the subspace basis $\{|00\rangle, |11\rangle\}$. The choice of $\theta = \pi/4$ yields the Bell state B_+_a maximally entangled state. It is the quantum entanglement of states that will give rise to the exponential speed-up of a quantum algorithm. The QLA is a sequence of interleaved unitary collision-streaming operators that entangle the qubits and then spread that entanglement throughout the lattice.

2. The Dyson map and the generation of a unitary evolution equation for Maxwell equations

Consider the subset of Maxwell equations

$$\nabla \times \mathbf{E} = -\frac{\partial \mathbf{B}}{\partial t} \quad , \quad \nabla \times \mathbf{H} = \frac{\partial \mathbf{D}}{\partial t} \quad (4)$$

and treat $\nabla \cdot \mathbf{B} = 0$ and $\nabla \cdot \mathbf{D} = 0$ as initial constraints that remain satisfied in the continuum limit for all times. (This, of course, follows immediately from taking the divergence of Eq. (4)).

For lossless media, the electric and magnetic fields satisfy the constitutive relations for a tensor dielectric nonmagnetic medium

$$\mathbf{D} = \boldsymbol{\varepsilon} \cdot \mathbf{E}, \quad \mathbf{B} = \mu_0 \mathbf{H}. \quad (5)$$

For Hermitian $\boldsymbol{\varepsilon}$, one can transform to a coordinate system in which $\boldsymbol{\varepsilon}$ is diagonal. Eq. (5) can be rewritten in matrix form

$$\mathbf{d} = \mathbf{W}\mathbf{u} \quad \text{, with} \quad \mathbf{d} \doteq (\mathbf{D}, \mathbf{B})^T, \quad \mathbf{u} \doteq (\mathbf{E}, \mathbf{H})^T \quad (6)$$

where $\mathbf{W}$ is a 6×6 Hermitian block diagonal constitutive matrix

$$\mathbf{W} = \begin{bmatrix} \boldsymbol{\varepsilon}\mathbb{I}_{3\times3} & 0_{3\times3} \\ 0_{3\times3} & \mu_0\mathbb{I}_{3\times3} \end{bmatrix}. \quad (7)$$

$\mathbb{I}_{3\times3}$ is the 3×3 identity matrix, and the superscript $\mathbf{T}$ in Eq. (6) is the transpose. In matrix form, the Maxwell equations, Eq. (4) become

$$i\frac{\partial \mathbf{d}}{\partial t} = \mathbf{M}\mathbf{u} \tag{8}$$

where under standard boundary conditions, the curl-matrix operator $\mathbf{M}$ is Hermitian:

$$\mathbf{M} = \begin{bmatrix} 0_{3\times3} & i\nabla\times \\ -i\nabla\times & 0_{3\times3} \end{bmatrix}. \tag{9}$$

From Eq. (6), since $\mathbf{W}^{-1}$ exists, $\mathbf{u} = \mathbf{W}^{-1}\mathbf{d}$, so that Eq. (8) can be written

$$i\frac{\partial \mathbf{u}}{\partial t} = \mathbf{W}^{-1}\mathbf{M}\mathbf{u} \tag{10}$$

If the medium is homogeneous, then $\mathbf{W}^{-1}$ is constant and will commute with the curl-operator $\mathbf{M}$. Under these conditions, the product $\mathbf{W}^{-1}\mathbf{M}$ is Hermitian and Eq. (10) gives unitary evolution for $\mathbf{u} = (\mathbf{E}, \mathbf{H})^T$.

However, if the medium is spatially inhomogeneous, then $\left[\mathbf{W}^{-1}, \mathbf{M}\right] \neq 0$ and the evolution equation for the $\mathbf{u}$-field is not unitary.

2.1 Dyson map

To determine a unitary evolution of the electromagnetic fields in an inhomogeneous dielectric medium, it [20] has been shown that there exists a Dyson map ρ: $\mathbf{u} \rightarrow \mathbf{Q}$ such that in the new field variables $\mathbf{Q}$ the resulting evolution equation will be unitary. For the Maxwell equations consider

$$\mathbf{Q} = \rho\mathbf{u} = \mathbf{W}^{1/2}\mathbf{u}. \tag{11}$$

For time-independent media, the evolution equation for the new fields $\mathbf{Q}$ is

$$i\rho\frac{\partial \mathbf{u}}{\partial t} = \rho\mathbf{W}^{-1}\mathbf{M}\rho^{-1}\rho\mathbf{u} \quad \Rightarrow i\frac{\partial \mathbf{Q}}{\partial t} = \rho\mathbf{W}^{-1}\mathbf{M}\rho^{-1}\mathbf{Q} \tag{12}$$

and is indeed unitary. Explicitly, the new fields, Eq. (11) and the ρ are

$$\mathbf{Q} = \begin{bmatrix} q_0 \\ q_1 \\ q_2 \\ q_3 \\ q_4 \\ q_5 \end{bmatrix} = \begin{bmatrix} n_x E_x \\ n_y E_y \\ n_z E_z \\ \mu_0^{1/2} H_x \\ \mu_0^{1/2} H_y \\ \mu_0^{1/2} H_z \end{bmatrix}, \quad \rho = \begin{bmatrix} n_x & 0 & 0 & 0 & 0 & 0 \\ 0 & n_y & 0 & 0 & 0 & 0 \\ 0 & 0 & n_z & 0 & 0 & 0 \\ 0 & 0 & 0 & \mu_0^{1/2} & 0 & 0 \\ 0 & 0 & 0 & 0 & \mu_0^{1/2} & 0 \\ 0 & 0 & 0 & 0 & 0 & \mu_0^{1/2} \end{bmatrix} \tag{13}$$

The refractive index $n_i = \sqrt{\varepsilon_i}$. Typically, we will use units where $\mu_0 = 1$. In component form, Maxwell equations for fields and with constitutive matrix restricted to spatially 2D (x, y) dependence, Eq. (12) reduces to

$$\frac{\partial q_0}{\partial t} = \frac{1}{n_x}\frac{\partial q_5}{\partial y}, \qquad \frac{\partial q_1}{\partial t} = -\frac{1}{n_y}\frac{\partial q_5}{\partial x}, \qquad \frac{\partial q_2}{\partial t} = \frac{1}{n_z}\left[\frac{\partial q_4}{\partial x} - \frac{\partial q_3}{\partial y}\right]$$
$$\frac{\partial q_3}{\partial t} = -\frac{\partial (q_2/n_z)}{\partial y}, \qquad \frac{\partial q_4}{\partial t} = \frac{\partial (q_2/n_z)}{\partial x}, \qquad \frac{\partial q_5}{\partial t} = -\frac{\partial (q_1/n_y)}{\partial x} + \frac{\partial (q_0/n_x)}{\partial y} \tag{14}$$

3. A qubit lattice representation for 2D tensor dielectric media

QLA consists of a sequence of unitary collision and streaming operators on a 2D spatial lattice, which will recover the continuum Maxwell equations, Eq. (14) to second order in the spatial grid size, δ. In particular, we need to have 6 qubits/lattice sites to represent the field components in Eq. (13). QLA permits us to handle the x- and y-dependence separately. Let us first consider the x-dependence and recover the $\partial q_i/\partial x$ - terms. From Eq. (14), we see coupling between $q_1 \leftrightarrow q_5, q_2 \leftrightarrow q_4$, Hence, we introduce the local entangling collision operator

$$C_X = \begin{bmatrix} 1 & 0 & 0 & 0 & 0 & 0 \\ 0 & \cos\theta_1 & 0 & 0 & 0 & -\sin\theta_1 \\ 0 & 0 & \cos\theta_2 & 0 & -\sin\theta_2 & 0 \\ 0 & 0 & 0 & 1 & 0 & 0 \\ 0 & 0 & \sin\theta_2 & 0 & \cos\theta_2 & 0 \\ 0 & \sin\theta_1 & 0 & 0 & 0 & \cos\theta_1 \end{bmatrix} \tag{15}$$

The collision angles θ_1 and θ_2 need to be chosen to recover the refractive index factors before the corresponding spatial derivatives,

$$\theta_1 = \frac{\delta}{4n_y} \quad , \quad \theta_2 = \frac{\delta}{4n_z}. \tag{16}$$

The first of the unitary streaming operators will stream qubits q_1, q_4 one lattice unit in either direction while leaving the other four qubits fixed: $S_{14}^{\pm}$. The other unitary streaming operator will act on qubits q_2, q_5: $S_{25}^{\pm}$. The final unitary collide-stream sequence, $\mathbf{U_X}$ in the x-direction that leads to a second-order scheme in δ can be shown to be

$$\mathbf{U_X} = S_{25}^{+x}.C_X^{\dagger}.S_{25}^{-x}.C_X.S_{14}^{-x}.C_X^{\dagger}.S_{14}^{+x}.C_X.S_{25}^{-x}.C_X.S_{25}^{+x}.C_X^{\dagger}.S_{14}^{+x}.C_X.S_{14}^{-x}.C_X^{\dagger}. \tag{17}$$

It should be noted that if only applies the first 4 collide-stream sequence in Eq. (17) then the algorithm would only be first-order accurate.

Similarly, to recover the $\partial q_i/\partial y$ terms one would collisionally entangle qubits $q_0 \leftrightarrow q_5, q_2 \leftrightarrow q_3$ with

$$C_Y = \begin{bmatrix} \cos\theta_0 & 0 & 0 & 0 & 0 & \sin\theta_0 \\ 0 & 1 & 0 & 0 & 0 & 0 \\ 0 & 0 & \cos\theta_2 & \sin\theta_2 & 0 & 0 \\ 0 & 0 & -\sin\theta_2 & \cos\theta_2 & 0 & 0 \\ 0 & 0 & 0 & 0 & 1 & 0 \\ -\sin\theta_0 & 0 & 0 & 0 & 0 & \cos\theta_0 \end{bmatrix}, \tag{18}$$

with corresponding collision angles θ_0 and θ_2. θ_2 is given in Eq. (16), and

$$\theta_0 = \frac{\delta}{4n_x}. \tag{19}$$

The streaming operator S^y_{03} will act on qubits q_0, q_3 only and similarly for the operator S^y_{25}. The final unitary collide-stream second-order accurate or the y-direction for Maxwell equations is

$$\mathbf{U_Y} = S^{+y}_{25}.C^\dagger_Y.S^{-y}_{25}.C_Y.S^{-y}_{03}.C^\dagger_Y.S^{+y}_{03}.C_Y.S^{-y}_{25}.C_Y.S^{+y}_{25}.C^\dagger_Y.S^{+y}_{03}.C_Y.S^{-y}_{03}.C^\dagger_Y \tag{20}$$

We still need to recover the spatial derivatives on the refractive index components in Eq. (14). To obtain the $\partial n_z/\partial x$ and $\partial n_y/\partial x$ terms, we introduce the (non-unitary) sparse potential matrix

$$V_X = \begin{bmatrix} 1 & 0 & 0 & 0 & 0 & 0 \\ 0 & 1 & 0 & 0 & 0 & 0 \\ 0 & 0 & 1 & 0 & 0 & 0 \\ 0 & 0 & 0 & 1 & 0 & 0 \\ 0 & 0 & -\sin\beta_2 & 0 & \cos\beta_2 & 0 \\ 0 & \sin\beta_0 & 0 & 0 & 0 & \cos\beta_0 \end{bmatrix} \tag{21}$$

with collision angles

$$\beta_0 = \delta^2 \frac{\partial n_y/\partial x}{n_y^2} \quad , \quad \beta_2 = \delta^2 \frac{\partial n_z/\partial x}{n_z^2}, \tag{22}$$

while the corresponding (non-unitary) sparse potential matrix to recover the $\partial/\partial y$-derivatives in the refractive index components is

$$V_Y = \begin{bmatrix} 1 & 0 & 0 & 0 & 0 & o \\ 0 & 1 & 0 & 0 & 0 & 0 \\ 0 & 0 & 1 & 0 & 0 & 0 \\ 0 & 0 & \cos\beta_3 & \sin\beta_3 & 0 & 0 \\ 0 & 0 & 0 & 0 & 1 & 0 \\ -\sin\beta_1 & 0 & 0 & 0 & 0 & \cos\beta_1 \end{bmatrix} \tag{23}$$

with collision angles

$$\beta_1 = \delta^2 \frac{\partial n_x/\partial y}{n_x^2} \quad , \quad \beta_3 = \delta^2 \frac{\partial n_z/\partial y}{n_z^2}. \tag{24}$$

Thus, the final discrete QLA, that models the 2D Maxwell equations, Eq. (14), to $O(\delta^2)$, advances the lattice qubit-vector $\mathbf{Q}(t)$ to $\mathbf{Q}(t + \Delta t)$ is

$$\mathbf{Q}(t + \Delta t) = V_Y.V_X.\mathbf{U_Y}.\mathbf{U_X}.\mathbf{Q}(t) \tag{25}$$

Provided, we have diffusion ordering in the space–time lattice, that is, $\Delta t = \delta^2$. It is this ordering that requires us to have the unitary collision angles to be $O(\delta)$, Eqs. (16) and (19), and the external potential angles $O(\delta^2)$, Eqs. (22). We note that computationally QLA is more accurate if we employ the external potentials twice: once halfway through the collide-stream sequence and then at the end.

3.1 Non-unitary external potential operators

Recently, considerable effort has been expended into developing more efficient approximation for handling the evolution operator of a complex Hamiltonian system than the standard Suzuki-Trotter expansion Eqs. (21) and (23) [22]. In particular, the idea [23, 24] has been floated of approximating the full unitary operator by a *sum* of unitary operators. The actual implementation onto a quantum computer we will leave to another paper, as one of the outcomes of QLA discussed here will be a quantum-inspired highly efficient classical supercomputer algorithm. Moreover, its encoding onto a quantum computer will require error-correcting qubits with long coherence times, something currently out of reach in the noisy qubit regime we are in. Here, we will show the 4 unitary operators needed whose sum yields the sparse non-unitary potential operator V_X, Eq. (21). Letting $\mathbb{I}_6$ be the 6×6 identity matrix, then it is easily verified that

$$V_X = \frac{1}{2}\sum_{i=1}^{4} LCU_i \tag{26}$$

where the first two unitaries are diagonal

$$LCU_1 = \mathbb{I}_6 \quad , \quad LCU_2 = diag(-1,1,1,-1,-1,\ 1) \tag{27}$$

and the remaining two unitaries are

$$LCU_3 = \begin{bmatrix} 1 & 0 & 0 & 0 & 0 & o \\ 0 & \cos\beta_0 & 0 & 0 & 0 & -\sin\beta_0 \\ 0 & 0 & \cos\beta_2 & 0 & \sin\beta_2 & 0 \\ 0 & 0 & 0 & 1 & 0 & 0 \\ 0 & 0 & -\sin\beta_2 & 0 & \cos\beta_2 & 0 \\ 0 & \sin\beta_0 & 0 & 0 & 0 & \cos\beta_0 \end{bmatrix}, \tag{28}$$

and

$$LCU_4 = \begin{bmatrix} 1 & 0 & 0 & 0 & 0 & o \\ 0 & -\cos\beta_0 & 0 & 0 & 0 & \sin\beta_0 \\ 0 & 0 & -\cos\beta_2 & 0 & -\sin\beta_2 & 0 \\ 0 & 0 & 0 & 1 & 0 & 0 \\ 0 & 0 & -\sin\beta_2 & 0 & \cos\beta_2 & 0 \\ 0 & \sin\beta_0 & 0 & 0 & 0 & \cos\beta_0 \end{bmatrix}. \tag{29}$$

3.2 Conservation of energy

In a fully unitary representation, the norm of **Q** is a constant of the motion. This is simply the conservation of energy in the electromagnetic field. For fields being a function of (x, y), we have from Eqs. (11)–(13)

$$\mathcal{E}(t) = \frac{1}{L^2}\int_0^L dxdy\mathbf{Q}\cdot\mathbf{Q} = \frac{1}{L^2}\int_0^L dxdy\left[\varepsilon_x E_x^2 + \varepsilon_y E_y^2 + \varepsilon_z E_z^2 + \mu_o\left(B_x^2 + B_y^2 + B_z^2\right)\right] \quad (30)$$

where the (diagonal) tensor dielectric $\varepsilon_x = n_x^2$, ... and we restrict ourselves to nonmagnetic materials, for simplicity.

4. 2D numerical simulations from the QLA for electromagnetic scattering from 2D dielectric objects

We now present detailed QLA simulations of the initial value problem of the scattering of a 1D electromagnetic pulse from a localized dielectric object. In particular, we consider a 1D Gaussian pulse propagating in the x-direction toward a localized dielectric object of refractive index $n(x, y)$. The initial pulse has nonzero field components E_z, B_y, **Figure 1**, and scatters from either a localized cylindrical dielectric, **Figure 2a**, or a conic dielectric object, **Figure 2b**. These simulations were performed for $\delta = 0.1$.

It should be noted that QLA is an initial value algorithm. The refractive index profiles are smooth (e.g., hyperbolic tangents for the dielectric cylinder with boundary layer thickness ≈10 lattice units) and so *no* internal boundary conditions are imposed at any time in the simulation.

4.1 Effects of broken symmetry

When the 1D pulse scatters off the dielectric object with refractive index $n(x, y)$ the initial electric field spatial dependence $E_z(x)$ now becomes a function $E_z(x, y)$, while the initial magnetic field $B_y(x)$ will become a function $B_y(x, y)$. Now the scattered field

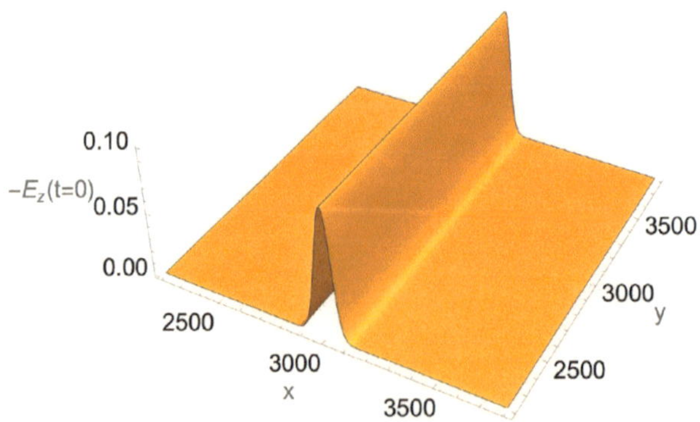

Figure 1.
A 1D electromagnetic pulse with initial fields $-E_z(x, t = 0), B_y(x, t = 0)$. 2D simulation grid $L \times L$ with $L = 8192$. Pulse full-width (in lattice units) ≈ 200. Since the Maxwell equations are linear and homogenous, the initial amplitude of the fields is arbitrary.

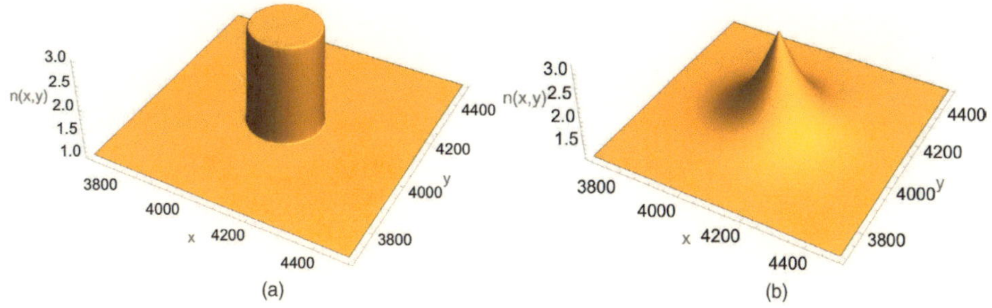

Figure 2.
(a) The dielectric cylinder, diameter ≈ 200, has rapidly increasing boundary dielectric from vacuum $n = 1$ to $n_{max} = 3$, whereas (b) the conic dielectric, base ≈ 240, has smoothly increasing dielectric from vacuum to conic peak of $n_{max} = 3$.

has $\partial B_y/\partial y \neq 0$. Thus, for $\nabla \cdot \mathbf{B} = 0$, the scattered field must develop an appropriate $B_x(x, y)$. This is seen in our QLA simulations, even though the explicit discrete collide-stream algorithm only models asymptotically the Maxwell subset, Eq. (4). $\nabla \cdot \mathbf{D} = 0$ and this is exactly conserved in our QLA simulation.

Similarly, for initial $E_y(x)$ polarization. In this case, the scattered magnetic field $B_z(x, y)$ satisfies, for our 2D scattering in the x-y plane, $\nabla \cdot \mathbf{B} = 0$ exactly and no other magnetic field components are generated. Our discrete QLA recovers this $\nabla \cdot \mathbf{B} = 0$ exactly. However, in an attempt to preserve $\nabla \cdot \mathbf{D} = 0$, the QLA will generate a nonzero $E_x(x, y)$ field.

4.2 Scattering from localized 2D dielectric objects with refractive index $n(x, y)$

Consider a 1D pulse with polarization $E_z(x, t)$ propagating in a vacuum toward a 2D dielectric scatterer. In **Figures 3–6**, we consider the time evolution of the resultant scattered E_z-field. The initial pulse is followed for a short time while it is propagating in the vacuum to verify that the QLA correctly determines its motion. As part of the pulse interacts with the dielectric object, the pulse speed within the dielectric itself is

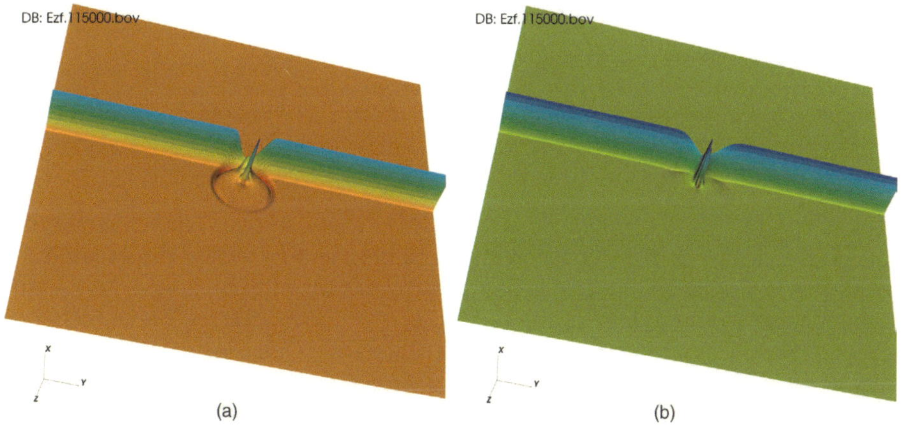

Figure 3.
The scattered E_z field after 15,000 iterations (i.e., $t = 15k$). (a) There is an internal reflection at the back of the cylindrical dielectric.

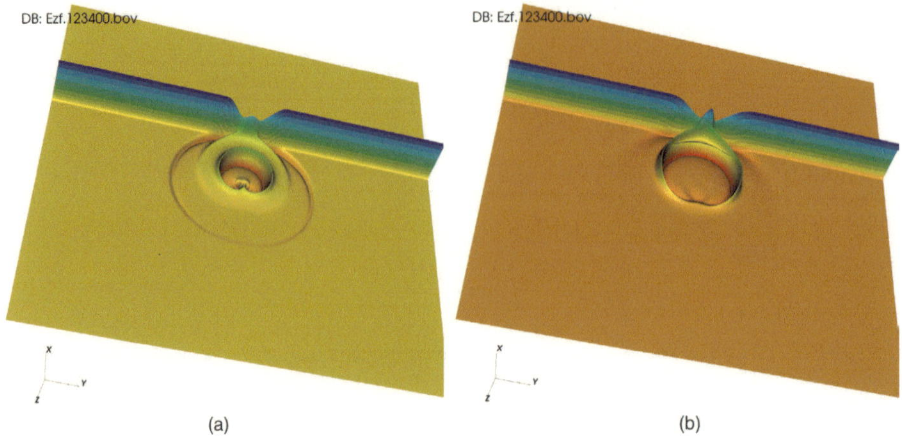

Figure 4.
The scattered E_z field at $t = 23.4k$. (a) a reflected circular wavefront occurs as that part of the pulse reaches the back-end of the cylindrical dielectric, along with the initial reflected circular wavefront with its π-changed phase at the front of the vacuum-cylinder boundary. (b) for the conic dielectric, there is an internal reflection from the apex of the cone's n_{max}, which then propagates out of the weakly varying cone edges.

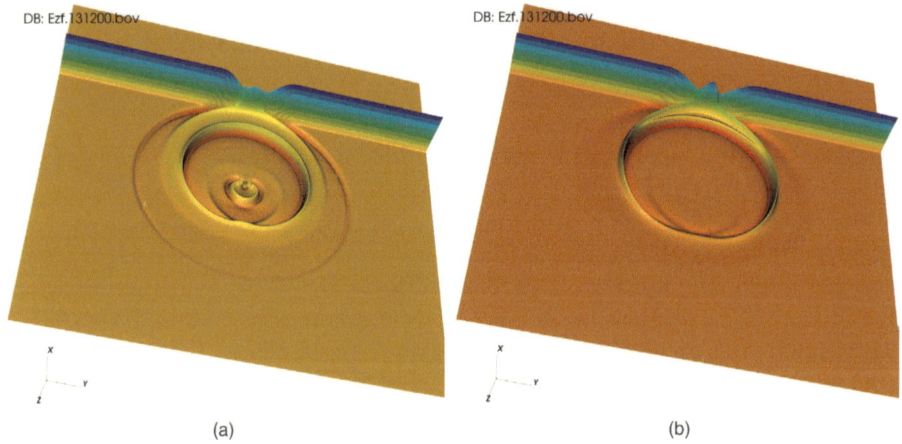

Figure 5.
The scattered E_z field at $t = 31.2k$. (a) There are multiple reflections/transmissions within the boundaries of the dielectric cylinder. (b) There is only one major reflection from the apex of the cone and which then propagates readily out from the cone edge.

decreased by the inverse of the refractive index profiles, $n(x, y)$. The remainder of the 1D pulse propagates undisturbed since it is still propagating within the vacuum.

One sees in **Figure 3a**, a circular-like wavefront reflecting back into the vacuum, with its E_z field π out of phase as the reflection is occurring from a low to higher refractive index around the vacuum-cylinder interface. One does not find such a reflected wavefront when the pulse interacts with the conic dielectric, **Figure 3b**.

At $t = 23.4k$, there is a major wavefront emanating from the back of the cylindrical dielectric, **Figure 4a**. For the conic dielectric, there is a major wavefront reflected from the apex of the conic dielectric, and this propagates out of the cone with a little reflection, **Figure 4b**.

DOI: http://dx.doi.org/10.5772/intechopen.112692

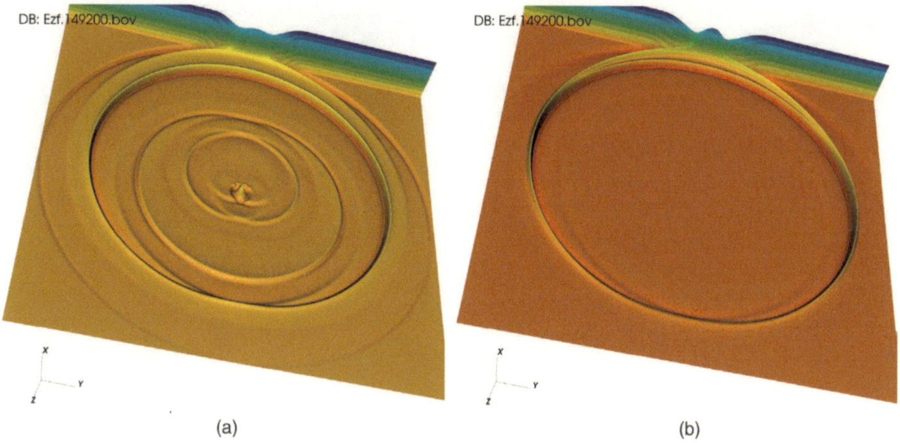

Figure 6.
The scattered E_z field at $t = 49.2k$. (a) There are multiple reflections/transmissions within the boundaries of the dielectric cylinder. (b) There is only one major reflection from the apex of the cone and which then propagates readily out from the cone edge.

One clearly sees at $t = 31.2k$ that more E_z wavefronts are being created because of the large refractive index gradients at the vacuum-cylinder dielectric boundary, while such gradients are missing from the vacuum-cone interface which leads to no new wavefronts in the scattering of the dielectric cone, **Figure** 5.

At $t = 49.2k$, the complex E_z wavefronts are due to repeated reflections and transmissions from the cylinder dielectric, **Figure 5a**. However, because of the slowly changing boundaries of the dielectric cone there are no more reflections and one sees only the outgoing wavefront from the pulses' interaction with the region around the n_{max} of the cone. Since the pulse reaches the apex of the cone before the corresponding pulse hits the backend of the dielectric cylinder, the conic wavefront is further advanced than that of the cylindrical wavefront, **Figure 5b**.

4.2.1 Auxiliary fields and $\nabla \cdot \mathbf{B}$

For incident E_z polarization and with 2D refractive index $n(x, y)$, the scattered electromagnetic fields will need to generate a B_x field in order to have $\nabla \cdot \mathbf{B} = 0$. In **Figure 6a** and **b**, we plot the self-generated $B_x(x, y)$ field at $t = 23.4k$ and $t = 49.2k$ for scattering from the dielectric cylinder. It is also found that $|\nabla \cdot \mathbf{B}|/B_0$ is typically zero everywhere in the spatial lattice except for a very localized region around the vacuum-dielectric boundary layer where the normalized $max\,(|\nabla \cdot \mathbf{B}|)/B_0$ reaches around 0.01 at very few isolated grid points (**Figure** 7).

4.3 Time dependence of (t) on perturbation parameter δ

The discrete total electromagnetic energy $\mathscr{E}(t)$. Eq. (30), is not constant since our current QLA is not totally unitary. However, the variations in $\mathscr{E}$ decrease significantly as $\delta \to 0$. δ is a measure of the discrete lattice spacing. The maximal variations occur shortly after the 1D pulse scatters from the 2D dielectric object. For $\delta = 0.1$, this occurs around $t = 15k$, with variations in the 5th decimal, Eq. (31). However, when

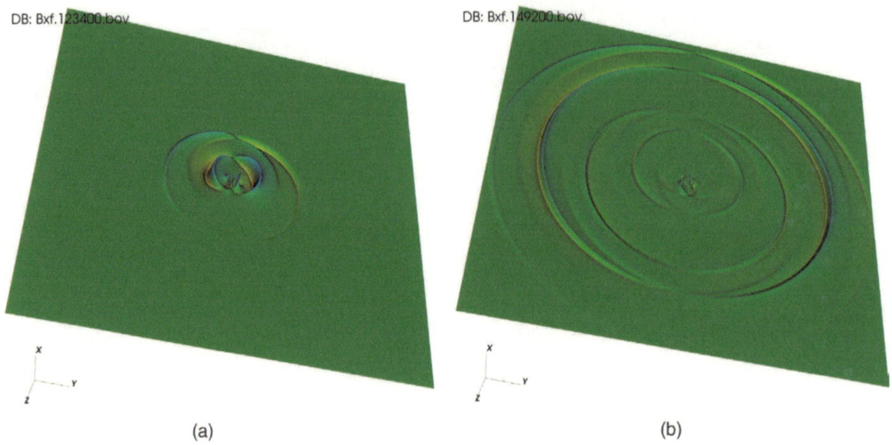

Figure 7.
The self-consistently generated $B_x(x, y)$-field after the 1D incident pulse with $B_y = B_y(x)$ scatters from a local dielectric with refractive index $n(x, y)$ at times: (a) $t = 23.4k$, corresponding to ***Figure 4a*** *for E_z, and (b) $t = 49.2k$, corresponding to* ***Figure 5a*** *for E_z.*

one reduces δ by a factor of 10 on the same lattice grid, then one recovers the same physics a factor of 10 later in time since δ controls the speed of propagation in the vacuum. Thus, the wallclock time of a QLA run is also increased by this factor of 10. We find, for $\delta = 0.01$, that the largest deviation in the total electromagnetic energy is now in the 8th decimal, Eq. (31).

$$\begin{array}{lcc} & \textit{time} & \mathscr{E}(t) \times 10^{-4} \\ \boldsymbol{\delta = 0.1}: & 0 & 1.442426615 \\ & 15000 & 1.442448\ldots \end{array} \qquad \begin{array}{lcc} & \textit{time} & \mathscr{E}(t) \times 10^{-4} \\ \boldsymbol{\delta = 0.01}: & 0 & 1.442426615 \\ & 150000 & 1.442426620 \end{array} \tag{31}$$

For $\delta = 10^{-3}$, there is variation in $\mathscr{E}$ in the 11th decimal.

4.4 Multiple reflections/transmissions within dielectric cylinder

We now examine the scattered electromagnetic fields—particularly the polarization $E_z(x, y)$—within and in the vicinity of the dielectric cylinder. These plots complement the global scattered $E_z(x, y)$ in **Figures 3a**, **4a**, and **5a**, but for better resolution we choose $\delta = 0.01$ and a slightly different ratio of pulse width to dielectric cylinder diameter. In **Figures 8–13**, the perspective is looking down from above with the 1D pulse propagating from left to right ($\rightarrow$), seen as a dark vertical band. The dielectric cylinder appears as a pink cylinder with the smaller darker pink being the base of the cylinder. The time is expressed in normalized time: $t = t^*/10$, where t^* is the QLA time for $\delta = 0.01$.

In **Figure 8a**, at time $t = 7.2k$, a part of the incident pulse has just entered the dielectric cylinder with the transmitted E_z field starting to lag behind the main 1D vacuum pulse since $1 < n_{cyl}$. Also, the reflected part of E_z emanates from the two boundary points at the sharp vacuum-dielectric boundary and has undergone a

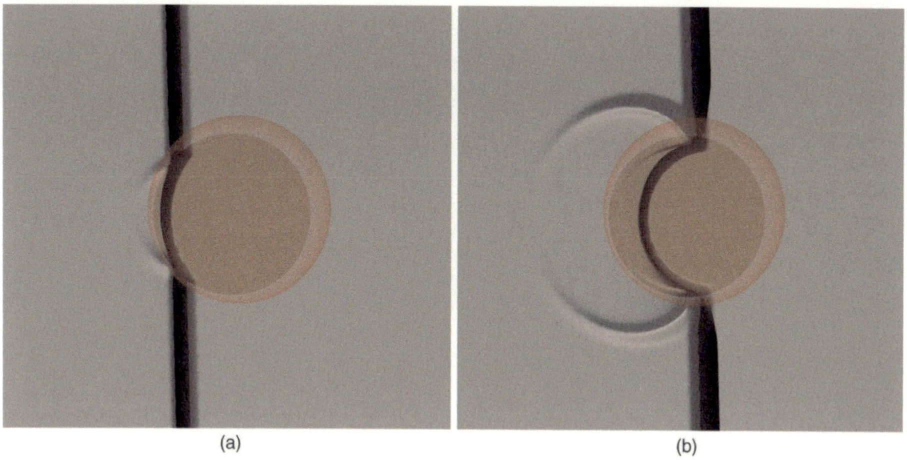

Figure 8.
A view from the z-axis of the 1D incident E_z wavefront, with x-y the plane of the page. The vacuum pulse is propagating in the x-direction, $\rightarrow$ (a) the 1D incident pulse has encountered the localized dielectric cylinder, with both transmission and reflection at the thin vacuum-dielectric boundary layer. The reflected E_z circular wavefront undergoes a π-phase change. (b) the transmitted E_z, within the dielectric, has a lower phase speed and so lags the 1D vacuum pulse.

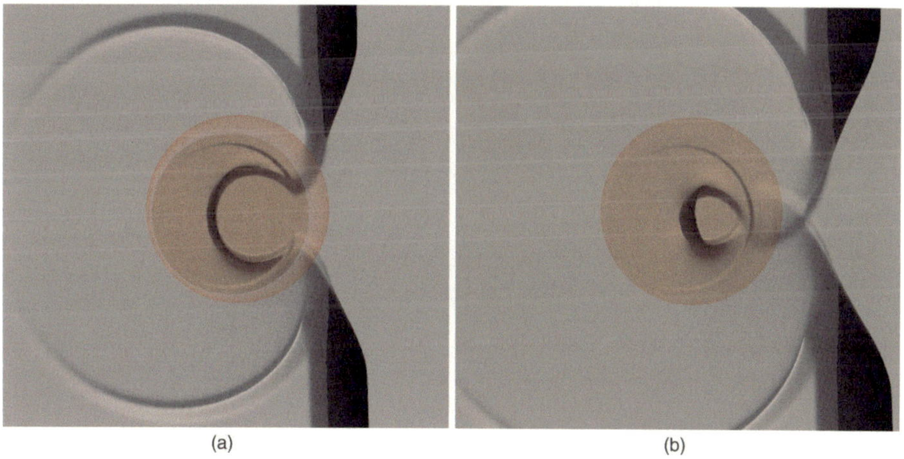

Figure 9.
As the 1D vacuum part of the wavefront moves past the dielectric cylinder, the two vacuum-dielectric boundary "points" move closer together: (a) at t = 18 k, (b) at t = 22.2 K. the vacuum-reflected wavefront keeps radiating out.

π-phase change because the incident 1D pulse is propagating from low to high refractive index. In **Figure 8b**, the slower transmitted E_z wavefront within the dielectric is very evident, as is the reflected part of E_z back into the vacuum.

By $t = 18k$, **Figure 7b**, the 1D pulse has propagated past the dielectric. The E_z-field within the dielectric is now being focussed due to its motion toward the backend of the cylinder, with its increasing amplitude but reduced base. As it reaches the backend of the dielectric, part of E_z will be transmitted into the vacuum while the other part will be reflected back into the dielectric but now without any phase change since the pulse is propagating from high to low refractive index.

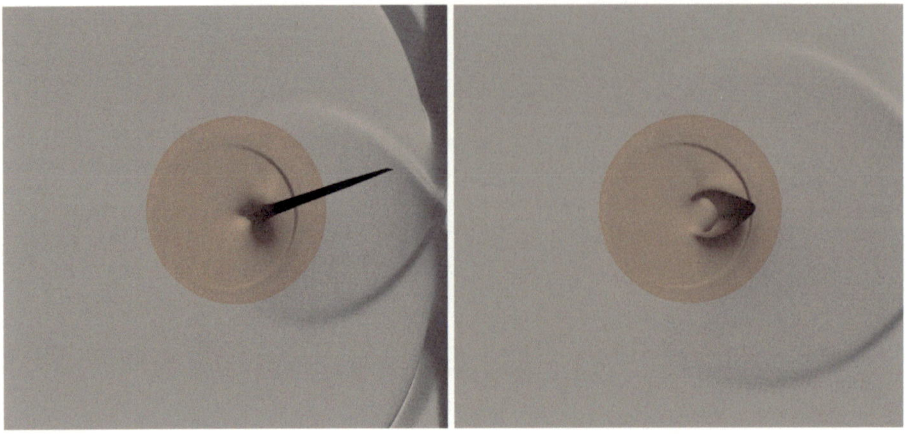

(a) (b)

Figure 10.
Wavefronts of E_z at times (a) $t = 28.2\ k$, and (b) $t = 32\ k$ around and within the dielectric cylinder after the original 1D pulse has moved past the dielectric. (a) the pinching of the two boundary "points" results in a focussing of E_z and its subsequent spiking at $t = 28.2k$. This spike now propagates toward the backend of the dielectric cylinder, (b), and "diffuses," one should also note the wavefront emanating from the 1D vacuum pulse.

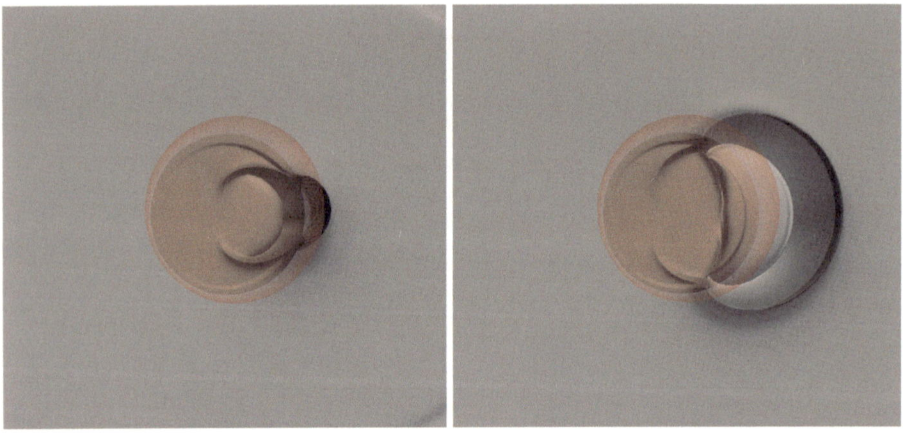

(a) (b)

Figure 11.
Wavefronts of E_z at times around and within the dielectric cylinder. At (a) $t = 38.4\ k$ the transmitted E_z within the dielectric is radiating outward, with one part reaching the back of the dielectric and resulting in a complex transmission into the vacuum region at the back end of the dielectric, (b) at $t = 43.2\ k$, and a complex reflection back into the dielectric. There is no phase change in the reflected E_z.

5. Dissipative classical systems, open quantum systems, and Kraus representation

So far we have treated Maxwell equations as a closed system based on the energy conservation dictated from the Hermiticity and positive definiteness of the constitutive matrix **W**, Eq. (7) since we have restricted ourselves to perfect materials. However, when we wish to consider actual materials, there is dissipation. This immediately defeats any attempt to pursue a unitary representation in the original Hilbert space. The obvious question is: can we embed our dissipative system into a higher dimension

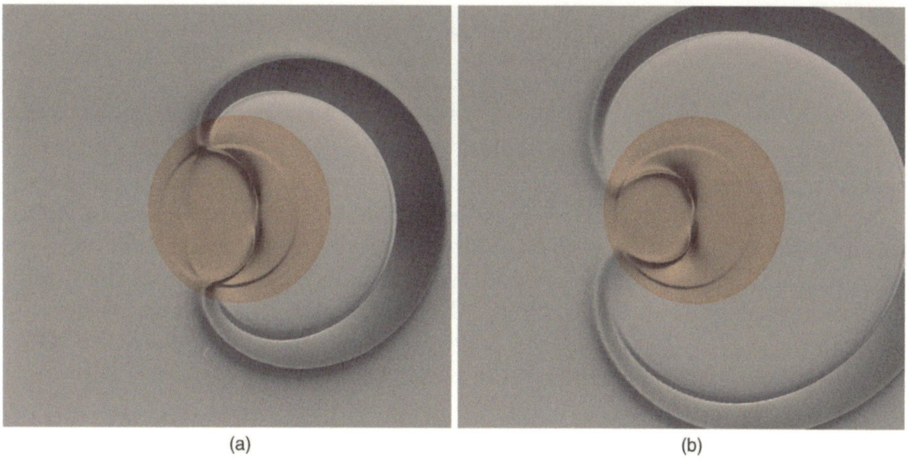

Figure 12.
Wavefronts of E_z at times (a) t = 47.4 k, and (b) t = 52.2 k around and within the dielectric cylinder. The major vacuum wavefront that is transmitted out of the dielectric now radiates out in the xy-plane. The two boundary contact "points" of the wavefront are now propagating back to the front of the dielectric cylinder, as clearly seen in (a) and (b). These localized wavefronts will have their global wavefronts similar to those shown in ***Figures 3a, 4a****, and* ***5a****.*

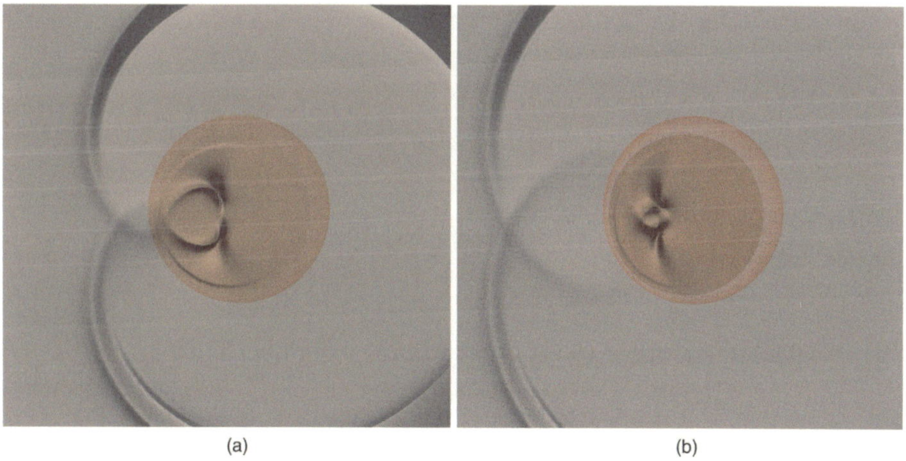

Figure 13.
Wavefronts of E_z at times (a) t = 55.8 k, and (b) t = 60 k around and within the dielectric cylinder.

closed Hilbert space, and thus recover unitary evolution in this new space and build an appropriate QLA that can be encoded onto quantum computers? To accomplish this, we resort to open quantum system theory [21] to describe classical dissipation as the observable result of interaction between our system of interest and its environment.

For a closed quantum system, the time evolution of a pure state $|\psi(t)\rangle$ is given by the unitary evolution from the Schrodinger equation: $|\psi(t)\rangle = U(t)|\psi(0)\rangle$ with $U = exp\left[-itH_0\right]$ unitary for the Hermitian Hamiltonian H_0. The evolution of the density matrix, $\rho = |\psi\rangle\langle\psi|$, is governed by the corresponding von Neumann equation: $\rho(t) = U(t)\rho(0)U^{\dagger}(t)$. The density matrix formulation is required when dealing with composite systems. Kraus realized that the density matrix retains its needed properties if one generalized its evolution operator to

$$\rho(t) = \sum_k K_k \rho(0) K_k^\dagger, \quad \text{with} \sum_k K_k^\dagger K_k = I \tag{32}$$

where the only restriction on the set of so-called Kraus matrices K_k is that the sum of $K_k^\dagger K_k$ is the identity matrix. The evolution of the density matrix, Eq. (32), is no longer unitary for $k \geq 2$.

The Kraus representation [21] is most useful when dealing with quantum noisy operations due to interaction with an environment. For those problems in which this noisy operation translates into a dissipative process, the Hamiltonian for the system in the Schrodinger representation has both a Hermitian part, H_0, and an anti-Hermitian part, iH_1, that models the dissipation. A simple but nontrivial example is the 1*D* Maxwell equations (without sources) for a homogeneous scalar medium with electrical losses,

$$i\frac{\partial}{\partial t}\begin{bmatrix} E_y \\ H_z \end{bmatrix} = \begin{bmatrix} 0 & \dfrac{\varepsilon^*}{|\varepsilon|^2}\hat{p}_x \\ \dfrac{1}{\mu_0}\hat{p}_x & 0 \end{bmatrix}\begin{bmatrix} E_y \\ H_z \end{bmatrix} \tag{33}$$

with complex permittivity $\varepsilon = \varepsilon_R + i\varepsilon_I$. $\varepsilon^* = \varepsilon_R - i\varepsilon_I$. $\hat{p}_x = -i\partial_x$ is the momentum operator. Introducing the Dyson map $\boldsymbol{\rho} = diag\left(|\varepsilon|/\sqrt{\varepsilon_R}, \sqrt{\mu_0}\right)$ into Eq. (33) and after some algebraic manipulations the evolution equation can be written as

$$i\frac{\partial \mathbf{Q}}{\partial t} = \left[v_\delta\left(\sigma_x + \frac{1}{2}\delta\sigma_y\right)\hat{p}_x - \frac{i}{2}\delta v_\delta \sigma_x \hat{p}_x\right]\mathbf{Q}, \tag{34}$$

where the state vector $\mathbf{Q} = \boldsymbol{\rho}\mathbf{u}$, where $\mathbf{u}$ is defined in Eq. (6). $\delta = \varepsilon_I/\varepsilon_R$ is the loss angle, v_δ is the phase velocity $v_\delta = 1/\sqrt{\varepsilon_R \mu_0 (1+\delta^2)}$ and $\sigma_x, \sigma_y, \sigma_z$ are the Pauli matrices.

5.1 Classical dissipation as a quantum amplitude damping channel

In symbolic form, the Maxwell equations with electric resistive losses, Eq. (34), can be written in the Schrodinger-form

$$i\frac{\partial|\psi_S\rangle}{\partial t} = \left(\hat{H}_0(\mathbf{r}) - i\hat{H}_1(\mathbf{r})\right)|\psi_S\rangle \tag{35}$$

where the Hamiltonians $\hat{H}_0$ and $\hat{H}_1$ are Hermitian, and the dissipative operator $i\hat{H}_1$ is anti-Hermitian and positive definite. The positive definiteness requirement for the specific case of propagation in a lossy medium translates to

$$Im\left[E_y^*\frac{\partial H_z}{\partial x} + H_z^*\frac{\partial E_y}{\partial x}\right] > 0, \quad \text{with} \quad \varepsilon_I > 0. \tag{36}$$

In general, the dissipative operator $\hat{H}_1$ is relatively simple and models the phenomenological or coarse-graining of the underlying microscopic dissipative processes.

We aim to represent the dissipation in the Schrodinger picture, Eq. (35), as an open quantum system S interacting with its environment *Env*. The full system,

$S + Env$, is closed, and hence its time evolution is unitary. Let $\hat{\mathcal{U}}$ be this unitary operator, and ρ the total density matrix with $\hat{\mathcal{U}} : \rho(0) \to \rho(t)$. We make the usual assumption that the initial total density matrix is separable into the system and into the environmental Hilbert spaces: $\rho(0) = \rho_S(0) \otimes \rho_E(0)$. A quantum operation E on the open system of interest is defined as the map that propagates the open system density in time t:

$$\rho_S(t) = E(\rho_S(0)). \tag{37}$$

But under the conditions of initial separability, the action of the full unitary operators on the total density matrix will yield, after taking the trace over the environment,

$$\rho_S(t) = Tr_E(\rho(t)) = Tr_E\left(\hat{\mathcal{U}}\rho(0)\hat{\mathcal{U}}^\dagger\right) \tag{38}$$

Assuming a stationary environment, $\rho_E(0) = |a\rangle\langle a|$, Eq. (38) can be written as

$$\rho_S(t) = \sum_\mu \hat{K}_\mu \rho_S(0) \hat{K}_\mu^\dagger. \tag{39}$$

where $\hat{K}_\mu = \langle \mu|\hat{\mathcal{U}}|a\rangle$. These operators $\hat{K}_\mu$ will form a Kraus representation for the quantum operation E for an open system, Eq. (37), provided the so-called Kraus operators satisfy the extended "unitarity" condition

$$\sum_\mu K_\mu^\dagger K_\mu = I \tag{40}$$

Note that the individual Kraus operator need not be unitary. Based on this framework for open quantum systems, we proceed to construct a physical unitary dilation for the combined system-environment by identifying dissipation as an amplitude-damping operation, [21].

Let d be the dimension of the system Hilbert space, and r the dimension of the dissipative Hamiltonian H_1, Eq. (35). We require $d \geq 2r$, for optimal results but the dilation technique can be also applied to systems with $d = r$. If the system was quantum mechanical in nature, then there can be a set of d^2 Kraus operators at most. The matrix representation of the total unitary dilation evolution operator consists of listing all the Kraus matrices in the first column block. The remaining columns must then be determined, so that $\hat{\mathcal{U}}$ is unitary. This unitary dilation is equivalent to the Stinespring dilation theorem [25]. The advantage of the Kraus approach is that it avoids the need to actually know the physical properties of the environment.

Returning to the Schrodinger representation of the classical system Eq. (35), one can employ the Trotter-Suzuki expansion to $exp\left[-i\delta t\left(\hat{H}_0 - i\hat{H}_1\right)\right]$

$$|\psi(\delta t)\rangle = \left[e^{-i\delta t\hat{H}_0} \cdot e^{-\delta t\hat{H}_1} + O\left(\delta t^2\right)\right]|\psi(0)\rangle. \tag{41}$$

Even though $exp\left(-\delta t\hat{H}_1\right)$ is not unitary, $\hat{H}_1$ is Hermitian and can be diagonalized by a unitary transformation U_1

$$\hat{H}_1 = \hat{U}_1\hat{D}_1\hat{U}_1^\dagger \quad \text{with diagonal} \quad \hat{D}_1 = diag[\gamma_1, \ldots, \gamma_r], \tag{42}$$

where $\gamma_i > 0$ are the dissipative rate eigenvalues of $\hat{H}_1$. Thus, Eq. (41) becomes

$$|\psi(\delta t)\rangle = \left[e^{-i\delta t \hat{H}_0}\hat{U}_1\hat{K}_0\hat{U}_1^\dagger + O(\delta t^2)\right]|\psi(0)\rangle, \tag{43}$$

where $\hat{K}_0$ is

$$\hat{K}_0 = \begin{bmatrix} \hat{\Gamma}_{r\times r} & 0_{r\times r} \\ 0_{(d-r)\times(d-r)} & I_{(d-r)\times(d-r)} \end{bmatrix}, \quad \text{with diagonal } \hat{\Gamma}_{r\times r} = diag\left[e^{-\gamma_1\delta t} \ldots e^{-\gamma_r\delta t}\right]. \tag{44}$$

The non-unitary $\hat{K}_0$ will be one of our Kraus operators, and it describes the physical dissipation in the open system. We must now introduce a second Kraus operator $\hat{K}_1$, so that $\hat{K}_0^\dagger\hat{K}_0 + \hat{K}_1^\dagger\hat{K}_1 = \mathcal{I}$:

$$\hat{K}_1 = \begin{bmatrix} 0_{(d-r)\times(d-r)} & 0_{(d-r)\times(d-r)} \\ \sqrt{I_{r\times r} - \hat{\Gamma}^2} & 0_{r\times r} \end{bmatrix}. \tag{45}$$

$\hat{K}_1$ represents a transition that is not of direct interest. These Kraus operators $\hat{K}_0, \hat{K}_1$ are the multidimensional analogs of the quantum amplitude damping channel [21]: with $\hat{K}_0$ corresponding to the dissipation processes, while $\hat{K}_1$ corresponds to an unwanted quantum transition.

The block structure of the final *unitary* dilation evolution operator $\hat{\mathcal{U}}_{diss}$, corresponding to the non-unitary dissipation operator $e^{-\delta t\hat{H}_1}$, consists of column blocks of the Kraus operators $(K_0 \quad K_1 \ldots)^T$, and the remaining column blocks are of those matrices required to make $\hat{\mathcal{U}}_{diss}$ unitary [21]:

$$\hat{\mathcal{U}}_{diss} = \begin{bmatrix} \hat{\Gamma} & 0 & 0 & -\sqrt{I_{r\times r} - \hat{\Gamma}^2} \\ 0 & I_{(d-r)\times(d-r)} & 0 & 0 \\ 0 & 0 & I_{(d-r)\times(d-r)} & 0 \\ \sqrt{I_{r\times r} - \hat{\Gamma}^2} & 0 & 0 & \hat{\Gamma} \end{bmatrix}. \tag{46}$$

Thus, it can be shown that the evolution of the system $|\psi_0\rangle$ and environment $|0\rangle$ is given by

$$|0\rangle|\psi_0\rangle = \frac{1}{\sqrt{E_0}}\sum_q \psi_{0q}|0\rangle|q\rangle \to |0\rangle \otimes e^{-i\delta t\hat{H}_0}\hat{U}_1\hat{K}_0\hat{U}_1^\dagger|\psi_0\rangle + |1\rangle \otimes \hat{K}_1|\psi_0\rangle, \tag{47}$$

where measurement of the first qubit-environment by $|0\rangle\langle 0| \otimes I_{d\times d}$ yields a state analogous to $|0\rangle|\psi(\delta t)\rangle$. Finally, on taking the trace over the environment will yield the desired system state $|\psi(\delta t)\rangle$. The corresponding quantum circuit for Eq. (47) is shown in **Figure 14.**

It is important to highlight that the implementation of the dissipative case is directly overlapping with the QLA framework. The QLA can be used to implement the $\exp\left[-i\delta t\hat{H}_0\right]$ part in Eq. (43) as proposed in the previous sections. Specifically, for the lossy medium, the exponential operator of the $\hat{H}_0$ term in Eq. (35) can be easily handled with QLA [12, 13].

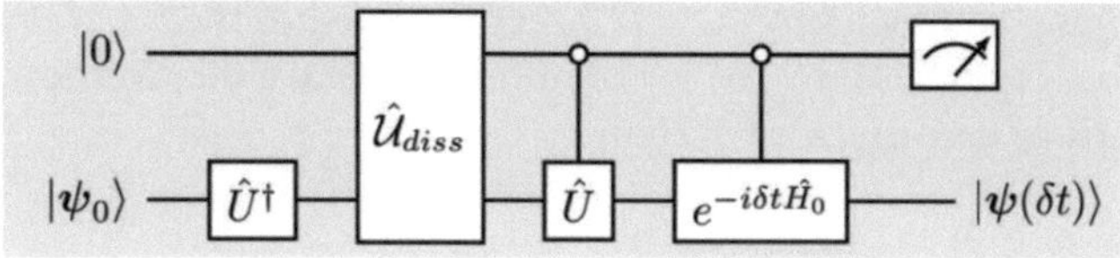

Figure 14.
Quantum circuit diagram for Eq. (47).

6. Conclusions

The Schrodinger-Dirac equations are the backbone of the work presented here on Maxwell equations both in lossless inhomogeneous and lossy dielectric media. In both cases, straightforward application of unitary algorithms fail, in the first case, somewhat surprisingly one finds that even though a Dyson map points to the required electromagnetic field variables in a tensor dielectric, its implementation has till now defied a fully unitary representation. Our current QLA approach requires some external non-unitary operators that recover the terms involving the spatial derivatives on the refractive indices of the medium. These sparse matrices can be modeled by the sum of a linear combination of unitaries (LCU), which can then be encoded onto a quantum computer [23, 24]. In the second case, handling dissipative systems immediately forces us to consider an open quantum system interacting with its environment. Typically, this forces us into a density matrix formulation and a clever introduction of what is known as Kraus operators [21, 25]. The beauty of the Kraus representation is that even though the system of interest is interacting with the environment, the Kraus operators do not need detailed information on the environment.

We presented detailed 2D scattering of a 1D electromagnetic pulse off localized dielectric objects. QLA is an initial value scheme. No internal boundary conditions are imposed at the vacuum-dielectrix interface. For dielectrics will large spatial gradients in the refractive index, QLA simulations show strong internal reflection/transmission within the dielectric object. These lead to quite complex time evolution of wavefronts from the dielectric objects. On the other hand, for weak spatial gradients in the refractive index, there are negligible reflections from the vacuum-dielectric interface. This is reminiscent of WKB-like effects in the ray tracing approximation.

In considering the dissipative counterpart, one must now include both the system and its environment in order to get a closed system with unitary representation. The Kraus operators are the most general scheme that will retain the properties of the density matrix in time. The probability of obtaining the desired non-unitary evolution of the open system after the measurement operator $\hat{P}_0 = |0\rangle\langle 0| \otimes I_{d\times d}$ is

$$p(0) = \sum_{i=1}^{r} e^{-2\gamma_i \delta t} |\psi_{i0}|^2 + \sum_{i=r+1}^{d} |\psi_{i0}|^2 \geq 1 + \left(e^{-2\gamma_{max}\delta t} - 1\right) \sum_{i=1}^{r} e^{-2\gamma_i \delta t} |\psi_{i0}|^2 \tag{48}$$

The form of the unitary operator $\hat{\mathcal{U}}_{diss}$, Eq. (42) implies that it can be decomposed into r two-level unitary rotations $\hat{R}_y(\theta_i)$ with $\cos(\theta_i/2) = e^{-\gamma_i \delta t}$. Then, the quantum circuit implementation of $\hat{\mathcal{U}}_{diss}$ requires $O\left(r log_2^2 d\right)$ CNOT - and $\hat{R}_y(\theta_i)$ quantum gates, so that there is an improvement in the circuit depth of $O\left(r/d^2\right)$. Our multidimensional amplitude damping channel approach is directly related to the Sz. Nagy dilation by a rotation. The Sz. Nagy dilation [26] is the minimal unitary dilation containing the original dissipative (non-unitary) system.

Acknowledgements

This research was partially supported by Department of Energy grants DE-SC0021647, DE-FG02-91ER-54109, DE-SC0021651, DE-SC0021857, and DE-SC0021653. This work has been carried out within the framework of the EUROfusion Consortium, funded by the European Union via the Euratom Research and Training Programme (Grant Agreement No. 101052200 - EUROfusion). Views and opinions expressed, however, are those of the authors only and do not necessarily reflect those of the European Union or the European Commission. Neither the European Union nor the European Commission can be held responsible for them. E. K. is supported by the Basic Research Program, NTUA, PEVE. K.H is supported by the National Program for Controlled Thermonuclear Fusion, Hellenic Republic. This research used resources of the National Energy Research Scientific Computing Center (NERSC), a U.S. Department of Energy Office of Science User Facility located at Lawrence Berkeley National Laboratory, operated under Contract No. DE-AC02-05CH11231 using NERSC award FES-ERCAP0020430.

Author details

George Vahala[1*†], Min Soe[2†], Efstratios Koukoutsis[3†], Kyriakos Hizanidis[3†], Linda Vahala[4†] and Abhay K. Ram[5†]

1 William & Mary, Williambsurg, USA

2 Rogers State University, Claremore, USA

3 National Technical University of Athens, Zographou, Greece

4 Old Dominion University, Norfolk, USA

5 Massachusetts Institute of Technology, Cambridge, USA

*Address all correspondence to: gvahala@gmail.com

† These authors contributed equally.

References

[1] Boghosian BM, Taylor W IV. Quantum lattice gas models for the many-body Schrodinger equation. International Journal of Modern Physics. 1997;**C8**:705-716

[2] Boghosian BM, Taylor W IV. Simulating quantum mechanics on a quantum computer. Physica D. 1998;**120**:30-42

[3] Yepez J, Boghosian BM. An efficient and accurate quantum lattice-gas model for the many-body Schroedinger wave equation. Computer Physics Communications. 2002;**146**:280-294

[4] Vahala G, Yepez J, Vahala L. Quantum lattice gas representation of some classical solitons. Physics Letters A. 2003;**310**:187-196

[5] Vahala G, Vahala L, Yepez J. Inelastic vector soliton collisions: A lattice-based quantum representation. Philosophical Transactions: Mathematical, Physical and Engineering Sciences, Royal Society. 2004;**362**:1677-1690

[6] Yepez J, Vahala G, Vahala L, Soe M. Superfluid turbulence from quantum kelvin waves to classical Kolmogorov cascades. Physical Review Letters. 2009; **103**:084501

[7] Vahala G, Yepez J, Vahala L, Soe M, Zhang B, Ziegeler S. Poincare recurrence and spectral cascades in three-dimensional quantum turbulence. Physical Review E. 2011;**84**:046713

[8] Zhang B, Vahala G, Vahala L, Soe M. Unitary quantum lattice algorithm for two dimensional quantum turbulence. Physical Review E. 2011;**84**:046701

[9] Vahala G, Zhang B, Yepez J, Vahala L, Soe M. Chpt 11., unitary qubit lattice gas representation of 2D and 3D quantum turbulence. In: Oh HW, editor. Advanced Fluid Dynamics. London, UK: InTech; 2012. pp. 239-272

[10] Vahala L, Soe M, Vahala G, Yepez J. Unitary qubit lattice algorithms for spin-1 Bose Einstein condensates. Radiation Effects and Defects in Solids. 2019;**174**: 46-55

[11] Vahala G, Vahala L, Soe M. Qubit unitary lattice algorithms for spin-2 Bose-Einstein condensates, I - theory and Pade initial conditions. Radiation Effects and Defects in Solids. 2020;**175**:102-112

[12] Vahala G, Soe M, Vahala L. Qubit unitary lattice algorithms for spin-2 Bose-Einstein condensates, II - vortex reconnection simulations and non-abelian vortices. Radiation Effects and Defects in Solids. 2020;**175**:113-119

[13] Palpacelli S, S. Succi "the quantum lattice Boltzmann equation: Recent developments". Computer Physics Communications. 2008;**4**:980-1007

[14] Succi S, Fillion-Gourdeau F, Palpacelli S. Quantum lattice Boltzmann is a quantum walk. EPJ Quantum Technology. 2015;**2**:12

[15] Vahala G, Vahala L, Soe M, Ram AK. Unitary quantum lattice simulations for Maxwell equations in vacuum and in dielectric media. Journal of Plasma Physics. 2020;**86**:905860518

[16] Vahala G, Hawthorne J, Vahala L, Ram AK, Soe M. Quantum lattice representation for the curl equations of Maxwell equations. Radiation Effects and Defects in Solids. 2022;**177**:85

[17] Vahala G, Soe M, Vahala L, Ram A, Koukoutsis E, Hizanidis K. Qubit lattice algorithm simulations of Maxwell's

equations for scattering from anisotropic dielectric objects. e-print arXiv: 2301.13601. 2023

[18] Oganesov A, Vahala G, Vahala L, Soe M. Effect of Fourier transform on the streaming in quantum lattice gas algorithms. Radiation Effects and Defects in Solids. 2018;**173**:169

[19] Khan SA, Jagannathan R. A new matrix representation of the Maxwell equations based on the Riemann-Silberstein-weber vector for a linear inhomogeneous medium. arXiv: 2205.09907. 2022

[20] Koukoutsis E, Hizanidis K, Ram AK, Vahala G. Dyson maps and unitary evolution for Maxwell equations in tensor dielectric media. Physical Review A. 2023;**107**:042215

[21] Nielsen MA, Chuang IL. Quantum Computation and Quantum Information. 10th ed. New York: Cambridge University Press; 2010

[22] Suzuki M. Generalized Trotter's formula and systematic approximants of exponential operators and inner derivatives with applications to many-body probloems. Communications in Mathematical Physics. 1976;**51**:183

[23] Childs AM, Wiebe N. Hamiltonian simulation using linear combinations of unitary operations. Quantum Information and Computation. 2012;**12**:901

[24] Childs AM, Kothari R, Somma RD. Quantum algorithm for Systems of Linear Equations with exponentially improved dependence on precision. SIAM Journal on Computing. 2017;**46**: 1920

[25] Stinespring WF. Positive functions on C^*-algebras. Proceedings of the American Mathematical Society. 1955; **6**:211

[26] Paulsen V. Completely Bounded Maps and Operator Algebras. New York: Cambridge University Press; 2003

Chapter 8

Blow-up Solutions to Nonlinear Schrödinger Equation with a Potential

Masaru Hamano and Masahiro Ikeda

Abstract

This is a sequel to the paper "Characterization of the ground state to the intercritical NLS with a linear potential by the virial functional" by the same authors. We continue to study the Cauchy problem for a nonlinear Schrödinger equation with a potential. In the previous chapter, we investigated some minimization problems and showed global existence of solutions to the equation with initial data, whose action is less than the value of minimization problems and positive virial functional. In particular, we saw that such solutions are bounded. In this chapter, we deal with solutions to the equation with initial data, whose virial functional is negative contrary to the previous paper and show that such solutions are unbounded.

Keywords: nonlinear Schrödinger equation, linear potential, standing wave, blow-up, grow-up, global existence

1. Introduction

In this chapter, we consider the Cauchy problem of the following nonlinear Schrödinger equation with a linear potential:

$$i\partial_t u + \Delta_V u = -|u|^{p-1}u, \quad (t,x) \in \mathbb{R} \times \mathbb{R}^d, \tag{1}$$

where $d \geq 1$, $1 < p < 2^* - 1$,

$$2^* := \begin{cases} \infty & \text{if } d \in \{1,2\}, \\ \dfrac{2d}{d-2} & \text{if } d \geq 3, \end{cases} \tag{2}$$

and $\Delta_V := \Delta - V = \sum_{j=1}^{d} \frac{\partial^2}{\partial x_j^2} - V$. In particular, we consider the Cauchy problem of Eq. (1) with initial condition

$$u(0,\cdot) = u_0 \in H^1(\mathbb{R}^d). \tag{3}$$

Eq. (1) with $V \in L^\infty(\mathbb{R}^d)$ is a model proposed to describe the local dynamics at a nucleation site (see [1]).

Eq. (1) is locally well-posed in the energy space $H^1(\mathbb{R}^d)$ under some assumptions, where Eq. (1) is called local well-posedness in $H^1(\mathbb{R}^d)$ if Eq. (1) satisfies all of the following conditions:

- There is uniqueness in $H^1(\mathbb{R}^d)$ for a solution to Eq. (1).
- For each $u_0 \in H^1(\mathbb{R}^d)$, there exists a solution to Eq. (1) with Eq. (3) defined on a maximal existence interval $(T_{\min}, T_{\max})$, where $T_{\max} = T_{\max}(u_0) \in (0, \infty]$ and $T_{\min} = T_{\min}(u_0) \in [-\infty, 0)$.
- There is the blow-up alternative. That is, if $T_{\max} < \infty$ (resp. $T_{\min} > -\infty$), then we have

$$\lim_{t \uparrow T_{\max}} \|u(t)\|_{H_x^1} = \infty \left(\text{resp.} \lim_{t \downarrow T_{\min}} \|u(t)\|_{H_x^1} = \infty \right). \tag{4}$$

- The solution depends on continuously on the initial condition. That is, if $u_{0,n} \to u_0$ in $H^1(\mathbb{R}^d)$, then for any closed interval $I \subset (T_{\min}, T_{\max})$, there exists $n_0 \in \mathbb{N}$ such that for any $n \geq n_0$, the solution u_n to Eq. (1) with $u_n(0,x) = u_{0,n}(x)$ is defined on $C_t(I; H^1(\mathbb{R}^d))$ and satisfies $u_n \to u$ in $C_t(I; H^1(\mathbb{R}^d))$ as $n \to \infty$, where u is the solution to Eq. (1) with $u(0,x) = u_0(x)$.

To state a local well-posedness result, we define the space

$$\mathcal{K}_0(\mathbb{R}^d) := \overline{\{f \in L^\infty(\mathbb{R}^d) : \text{supp} f \text{ is compact.}\}}^{\|\cdot\|_{\mathcal{K}}}, \tag{5}$$

where

$$\|f\|_{\mathcal{K}} := \sup_{x \in \mathbb{R}^d} \int_{\mathbb{R}^d} \frac{|f(y)|}{|x-y|^{d-2}} dy. \tag{6}$$

We note that

$$L^{\frac{d}{2}-\varepsilon}(\mathbb{R}^d) \cap L^{\frac{d}{2}+\varepsilon}(\mathbb{R}^d) \hookrightarrow L^{\frac{d}{2},1}(\mathbb{R}^d) \hookrightarrow \mathcal{K}(\mathbb{R}^d) := \{f : \|f\|_{\mathcal{K}} < \infty\} \tag{7}$$

for some $\varepsilon > 0$, where the space $L^{p,q}(\mathbb{R}^d)$ denotes the usual Lorentz space.

Theorem 1 (Local well-posedness, [2–4]) Let $d \geq 1$ and $1 < p < 2^* - 1$. If V satisfies one of the following, then Eq. (1) is locally well-posed in $H^1(\mathbb{R}^d)$.

- $V \in L^\eta(\mathbb{R}^d) + L^\infty(\mathbb{R}^d)$ for $\eta \geq 1$ if $d = 1$ and $\eta > \frac{d}{2}$ if $d \geq 2$,
- $\|V_-\|_{\mathcal{K}} < 4\pi$ and $V \in L^{\frac{3}{2}}(\mathbb{R}^3) \cap \mathcal{K}_0(\mathbb{R}^3)$, where $V_- := \min\{V(x), 0\}$.

Moreover, the solution u to Eq. (1) conserves its mass and energy with respect to time t, where they are defined as

$$\begin{aligned} &(\text{Mass}) \quad M[u(t)] := \|u(t)\|_{L_x^2}^2, \\ &(\text{Energy}) \quad E_V[u(t)] := \frac{1}{2}\|u(t)\|_{\dot{H}_x^1}^2 + \frac{1}{2}\int_{\mathbb{R}^d} V(x)|u(t,x)|^2 - \frac{1}{p+1}\|u(t)\|_{L_x^{p+1}}^{p+1}. \end{aligned} \tag{8}$$

We turn to time behaviors of the solution to Eq. (1). A solution to Eq. (1) has various kinds of time behaviors by the choice of initial data. For example, we can consider the following time behaviors.

- (Scattering) We say that the solution u to Eq. (1) scatters in positive time (resp. negative time) if $T_{\max} = \infty$ (resp. $T_{\min} = -\infty$) and there exists $\psi_+ \in H^1(\mathbb{R}^d)$ (resp. $\psi_- \in H^1(\mathbb{R}^d)$) such that

$$\lim_{t\to+\infty} \|u(t) - e^{it\Delta_V}\psi_+\|_{H_x^1} = 0 \quad \left(\text{resp.} \lim_{t\to-\infty} \|u(t) - e^{it\Delta_V}\psi_-\|_{H_x^1} = 0\right), \tag{9}$$

where $e^{it\Delta_V}f$ is a solution to the corresponding linear equation with Eq. (1)

$$i\partial_t u(t,x) + \Delta_V u(t,x) = 0, \quad u(0,x) = f(x). \tag{10}$$

We say that u scatters when u scatters in positive and negative time.

- (Blow-up) We say that the solution u to Eq. (1) blows up in positive time (resp. negative time) if $T_{\max} < \infty$ (resp. $T_{\min} > \ \ \infty$). We say that u blows up when u blows up in positive and negative time.

- (Grow-up) We say that the solution u to Eq. (1) grows up in positive time (resp. negative time) if $T_{\max} = \infty$ (resp. $T_{\min} = -\infty$) and

$$\limsup_{t\to\infty} \|u(t)\|_{H_x^1} = \infty, \quad \left(\text{resp.} \limsup_{t\to-\infty} \|u(t)\|_{H_x^1} = \infty\right). \tag{11}$$

We say that u grows up when u grows up in positive and negative time.

- (Standing wave) We say that the solution u to Eq. (1) is a standing wave if $u = e^{i\omega t}Q_{\omega,V}$ for some $\omega \in \mathbb{R}$, where $Q_{\omega,V}$ satisfies the elliptic equation

$$-\omega Q_{\omega,V} + \Delta_V Q_{\omega,V} = -\left|Q_{\omega,V}\right|^{p-1} Q_{\omega,V}. \tag{12}$$

In particular, $Q_{\omega,V}$ is ground state to Eq. (12) if

$$Q_{\omega,V} \in \{\phi \in \mathcal{A}_{\omega,V} : S_{\omega,V}(\phi) \le S_{\omega,V}(\psi) \text{ for any } \psi \in \mathcal{A}_{\omega,V}\} =: \mathcal{G}_{\omega,V}, \tag{13}$$

where $S_{\omega,V}(f) := \frac{\omega}{2}M[f] + E_V[f]$ (and)

$$\mathcal{A}_{\omega,V} := \{\psi \in H^1(\mathbb{R}^d)\backslash\{0\} : S'_{\omega,V}(\psi) = 0\}. \tag{14}$$

We know the following results (Theorems 2 and 3) for time behaviors of the solutions to Eq. (1). For related results, we also list [5–38].

Theorem 2 (Hong, [3]) Let $d = p = 3$, $u_0 \in H^1(\mathbb{R}^3)$, and $Q_{1,0} \in \mathcal{G}_{1,0}$. Suppose that V satisfies $V \in L^{\frac{3}{2}}(\mathbb{R}^3) \cap \mathcal{K}_0(\mathbb{R}^3)$, $V \geq 0$, $x \cdot \nabla V \in L^{\frac{3}{2}}(\mathbb{R}^3)$, and $x \cdot \nabla V \leq 0$. We also assume that

$$M[u_0]E_V[u_0] < M[Q_{1,0}]E_0[Q_{1,0}] \quad \text{and} \|u_0\|_{L^2}\|u_0\|_{\dot{H}^1_V} < \|Q_{1,0}\|_{L^2}\|Q_{1,0}\|_{\dot{H}^1}. \tag{15}$$

Then, the solution u to Eq. (1) with Eq. (3) scatters.

Theorem 3 (Hamano–Ikeda, [4]) Let $d = 3$, $\frac{7}{3} < p < 5$, $u_0 \in H^1(\mathbb{R}^3)$, and $Q_{1,0} \in \mathcal{G}_{1,0}$. Suppose that V satisfies $V \geq 0$ and $x \cdot \nabla V \in L^{\frac{3}{2}}(\mathbb{R}^3)$. We also assume that

$$M[u_0]^{\frac{1-s_c}{s_c}} E_V[u_0] < M[Q_{1,0}]^{\frac{1-s_c}{s_c}} E_0[Q_{1,0}], \tag{16}$$

where $s_c := \frac{d}{2} - \frac{2}{p-1}$.

1. (Scattering)

   If $V \in L^{\frac{3}{2}}(\mathbb{R}^3) \cap \mathcal{K}_0(\mathbb{R}^3)$, $x \cdot \nabla V \leq 0$, and

$$\|u_0\|_{L^2}^{\frac{1-s_c}{s_c}} \|u_0\|_{\dot{H}^1} < \|Q_{1,0}\|_{L^2}^{\frac{1-s_c}{s_c}} \|Q_{1,0}\|_{\dot{H}^1}, \tag{17}$$

   then $(T_{\min}, T_{\max}) = \mathbb{R}$, that is, exists globally in time. Moreover, if u_0 and V are radially symmetric, then u scatters.

2. (Blow-up or grow-up)

   If "$V \in L^{\frac{3}{2}}(\mathbb{R}^3) \cap \mathcal{K}_0(\mathbb{R}^3)$ or $V \in L^{\sigma}(\mathbb{R}^3)$ for some $\frac{3}{2} < \sigma \leq \infty$," $2V + x \cdot \nabla V \geq 0$, and

$$\|u_0\|_{L^2}^{\frac{1-s_c}{s_c}} \|u_0\|_{\dot{H}^1_V} > \|Q_{1,0}\|_{L^2}^{\frac{1-s_c}{s_c}} \|Q_{1,0}\|_{\dot{H}^1}, \tag{18}$$

   then u blows up or grows up. Furthermore, if one of the following holds:

   - "u_0 and V are radially symmetric," $x \cdot \nabla V \geq 0$, and $V \in L^{\infty}(\mathbb{R}^3)$,
   - $xu_0 \in L^2(\mathbb{R}^3)$,

then u blows up.

Remark 1 Mizutani [39] proved that for any $\psi \in H^1$, there exists $\phi_{\pm} \in H^1(\mathbb{R}^3)$ such that

$$\lim_{t \to \pm\infty} \|e^{it\Delta_V}\psi - e^{it\Delta}\phi_{\pm}\|_{H^1_x} = 0 \tag{19}$$

under the assumptions $V \in L^{\frac{3}{2}}(\mathbb{R}^3)$ and $V \geq 0$, where the double-sign corresponds. Combining this limit and scattering part in Theorem 3 (or Theorem 2), we can see that the nonlinear solution u to Eq. (1) approaches to a free solution $e^{it\Delta}\phi_{\pm}$ as $t \to \pm\infty$ for some $\phi_{\pm} \in H^1(\mathbb{R}^3)$.

We realize that there is no potential, which satisfies scattering and blow-up or grow-up parts in Theorem 1 at the same time. Indeed, if V satisfies $x \cdot \nabla V \leq 0$ and $2V + x \cdot \nabla V \geq 0$, then $V \notin L^{\frac{3}{2}}(\mathbb{R}^3)$. Then, we consider a minimization problem

$$n_{\omega,V} := \inf\{S_{\omega,V}(f) : f \in H^1(\mathbb{R}^d)\backslash\{0\},\ K_V(f) = 0\} \tag{20}$$

to get a potential V, which deduces scattering and blow-up or grow-up at the same time. It proved in [40] that the condition Eq. (16) can be rewritten as the following by using $n_{\omega,V}$.

Proposition 1 Let $d \geq 3$, $1 + \frac{4}{d} < p < 2^* - 1$, $f \in H^1(\mathbb{R}^d)$, and $Q_{1,0} \in \mathcal{G}_{1,0}$. Assume that V satisfies (A2) with $|\alpha| \leq 1$ and (A6) below. Then, the following two conditions are equivalent.

1. $M[f]^{\frac{1-s_c}{s_c}} E_V[f] < M[Q_{1,0}]^{\frac{1-s_c}{s_c}} E_0[Q_{1,0}]$,

2. There exists $\omega > 0$ such that $S_{\omega,V}(f) < n_{\omega,V}$.

Using $n_{\omega,V}$, we expect that if $S_{\omega,V}(u_0) < n_{\omega,V}$ and $K_V(u_0) \geq 0$, then the solution u scatters and if $S_{\omega,V}(u_0) < n_{\omega,V}$ and $K_V(u_0) < 0$, then the solution u blows up or grows up, where K_V is called virial functional and is defined as

$$\begin{aligned} K_V(f) &:= \frac{d}{d\lambda}\bigg|_{\lambda=0} S_{\omega,V}\left(e^{d\lambda} f\left(e^{2\lambda}\cdot\right)\right) \\ &= 2\|f\|_{\dot{H}^1}^2 - \int_{\mathbb{R}^d} (x \cdot \nabla V)|f(x)|^2 dx - \frac{(p-1)d}{p+1}\|f\|_{L^{p+1}}^{p+1}. \end{aligned} \tag{21}$$

It is well known that $K_V(u(t))$ denotes variance of the solution and if $xu_0 \in L^2(\mathbb{R}^d)$ then

$$K_V(u(t)) = \frac{1}{4} \cdot \frac{d^2}{dt^2}\|xu(t)\|_{L_x^2}^2 \tag{22}$$

for each $t \in (T_{\min}, T_{\max})$. We also consider a minimization problem $r_{\omega,V}$, which restricts $n_{\omega,V}$ to radial functions, that is,

$$r_{\omega,V} := \inf\{S_{\omega,V}(f) : f \in H^1_{\text{rad}}(\mathbb{R}^d)\backslash\{0\},\ K_V(f) = 0\} \tag{23}$$

and expect for radial initial data u_0 and radial potential V that if $S_{\omega,V}(u_0) < r_{\omega,V}$ and $K_V(u_0) \geq 0$, then the solution u scatters and if $S_{\omega,V}(u_0) < r_{\omega,V}$ and $K_V(u_0) < 0$, then the solution u blows up. For more general minimization problems

$$\begin{aligned} n_{\omega,V}^{\alpha,\beta} &:= \inf\left\{S_{\omega,V}(f) : f \in H^1(\mathbb{R}^d)\backslash\{0\},\ K_{\omega,V}^{\alpha,\beta}(f) = 0\right\}, \\ r_{\omega,V}^{\alpha,\beta} &:= \inf\left\{S_{\omega,V}(f) : f \in H^1_{\text{rad}}(\mathbb{R}^d)\backslash\{0\},\ K_{\omega,V}^{\alpha,\beta}(f) = 0\right\} \end{aligned} \tag{24}$$

with

$$\alpha > 0, \quad \beta \geq 0, \quad 2\alpha - d\beta \geq 0, \tag{25}$$

the authors showed in [40, 41] the following results (Theorems 4 and 5) Eq. (27), where the functional $K^{\alpha,\beta}_{\omega,V}$ is given as

$$K^{\alpha,\beta}_{\omega,V}(f) := \frac{d}{d\lambda}\bigg|_{\lambda=0} S_{\omega,V}\left(e^{\alpha\lambda} f\left(e^{\beta\lambda}\cdot\right)\right). \tag{26}$$

Here, we realize $n_{\omega,V} = n^{d,2}_{\omega,V}$, $r_{\omega,V} = r^{d,2}_{\omega,V}$, and $K_V = K^{d,2}_{\omega,V}$.
To state the results, we give the assumptions of the potential V: Let $\mathfrak{a} \in (\mathbb{N} \cup \{0\})^d$.

A1. $V \in L^{\frac{3}{2}}(\mathbb{R}^3) \cap \mathcal{K}_0(\mathbb{R}^3)$

A2. $x^{\mathfrak{a}}\partial^{\mathfrak{a}} V \in L^{\frac{d}{2}}(\mathbb{R}^d) + L^{\sigma}(\mathbb{R}^d)$ for some $\frac{d}{2} \leq \sigma < \infty$

A3. $x^{\mathfrak{a}}\partial^{\mathfrak{a}} V \in L^{\frac{d}{2}}(\mathbb{R}^d) + L^{\infty}(\mathbb{R}^d)$

A4. $x^{\mathfrak{a}}\partial^{\mathfrak{a}} V \in L^{\eta}(\mathbb{R}^d) + L^{\sigma}(\mathbb{R}^d)$ for some $\frac{d}{2} < \eta \leq \sigma < \infty$

A5. $x^{\mathfrak{a}}\partial^{\mathfrak{a}} V \in L^{\eta}(\mathbb{R}^d) + L^{\infty}(\mathbb{R}^d)$ for some $\frac{d}{2} < \eta < \infty$

A6. $V \geq 0, x \cdot \nabla V \leq 0, 2V + x \cdot \nabla V \geq 0$

A7. $V \geq 0, x \cdot \nabla V \leq 0, \omega \geq \omega_0$ for

$$\omega_0 := -\frac{1}{2}\text{ess inf}_{x \in \mathbb{R}^d}(2V + x \cdot \nabla V). \tag{27}$$

We note that the third inequality implies $2V + x \cdot \nabla V + 2\omega \geq 0$ a.e. $x \in \mathbb{R}^d$.
Theorem 4 Let $d \geq 3$ and $1 + \frac{4}{d} < p < 2^* - 1$.

- (Non-radial case) Let V satisfy (A2) with $|\mathfrak{a}| \leq 1$ and (A6). Then, for each (α, β) with Eq. (25) and $\omega > 0$, $n^{\alpha,\beta}_{\omega,V} = n^{\alpha,\beta}_{\omega,0}$ holds. Moreover, if $x \cdot \nabla V < 0$, then $n^{\alpha,\beta}_{\omega,V}$ is never attained.

- (Radial case) Let V satisfy (A3) with $|\mathfrak{a}| \leq 1$ and (A7). Let V be radially symmetric. Then, $r^{\alpha,\beta}_{\omega,V}$ is attained for each (α, β) with Eq. (25). Moreover, if V satisfies (A3) with $|\mathfrak{a}| \leq 2$ and $3x \cdot \nabla V + x\nabla^2 V x^T \leq 0$, then $\mathcal{M}^{\alpha,\beta}_{\omega,V,\text{rad}} = \mathcal{G}_{\omega,V,\text{rad}}$ holds, where $\nabla^2 V$ denotes the Hessian matrix of V,

$$\begin{aligned}
\mathcal{M}^{\alpha,\beta}_{\omega,V,\text{rad}} &:= \left\{\phi \in H^1_{\text{rad}}(\mathbb{R}^d) : S_{\omega,V}(\phi) = r^{\alpha,\beta}_{\omega,V},\ K^{\alpha,\beta}_{\omega,V}(\phi) = 0\right\}, \\
\mathcal{G}_{\omega,V,\text{rad}} &:= \left\{\phi \in \mathcal{A}_{\omega,V,\text{rad}} : S_{\omega,V}(\phi) \leq S_{\omega,V}(\psi) \text{ for any } \psi \in \mathcal{A}_{\omega,V,\text{rad}}\right\}, \\
\mathcal{A}_{\omega,V,\text{rad}} &:= \left\{\psi \in H^1_{\text{rad}}(\mathbb{R}^d)\backslash\{0\} : S'_{\omega,V}(\psi) = 0\right\}.
\end{aligned} \tag{28}$$

The inequality $n^{\alpha,\beta}_{\omega,V} \leq r^{\alpha,\beta}_{\omega,V}$ holds by their definitions and the attainability of $n^{\alpha,\beta}_{\omega,V}$ and $r^{\alpha,\beta}_{\omega,V}$ deduces the following corollary.

Corollary 1 Under the all assumptions of (Non-radial case) in Theorem 4, we have

$$n^{\alpha,\beta}_{\omega,V} < r^{\alpha,\beta}_{\omega,V}. \tag{29}$$

Remark 2 In the case of $V = 0$, it is well known that $n^{\alpha,\beta}_{\omega,0}$ and $r^{\alpha,\beta}_{\omega,0}$ are attained by $Q_{\omega,0} \in \mathcal{G}_{\omega,0}$. That is, $n^{\alpha,\beta}_{\omega,0} = r^{\alpha,\beta}_{\omega,0} = S_{\omega,0}(Q_{\omega,0})$ holds.

Then, we investigate global existence of a solution to time-dependent Eq. (1).

Theorem 5 (Global well-posedness in H^1) Let $d \geq 3$ and $1 + \frac{4}{d} < p < 2^* - 1$.

- (Non-radial case) Let $u_0 \in H^1(\mathbb{R}^d)$ and $Q_{\omega,0} \in \mathcal{G}_{\omega,0}$. Suppose that V satisfies "(A1) or (A4) with $|\mathfrak{a}| = 0$," (A2) with $|\mathfrak{a}| = 1$, and (A6). We also assume that there exist (α, β) satisfying Eq. (25) and $\omega > 0$ such that

$$S_{\omega,V}(u_0) < S_{\omega,0}(Q_{\omega,0}) \ \left(= n^{\alpha,\beta}_{\omega,V}\right), \quad K^{\alpha,\beta}_{\omega,V}(u_0) \geq 0. \tag{30}$$

  Then, the solution u to Eq. (1) with Eq. (3) exists globally in time. In particular, it follows that

$$\sup_{t\in\mathbb{R}} \|u(t)\|_{H^1_x} < \infty. \tag{31}$$

- (Radial case) Let $u_0 \in H^1_{\text{rad}}(\mathbb{R}^d)$ and $Q_{\omega,V} \in \mathcal{G}_{\omega,V,\text{rad}}$. Suppose that V is radially symmetric and satisfies "(A1) or (A5) with $|\mathfrak{a}| = 0$," (A3) with $|\mathfrak{a}| = 1, 2$, (A7), and $3x \cdot \nabla V + x\nabla^2 V x^T \leq 0$. If there exist (α, β) with Eq. (25) and $\omega > 0$ satisfying $\omega \geq \omega_0$ such that

$$S_{\omega,V}(u_0) < S_{\omega,V}(Q_{\omega,V}) \ \left(= r^{\alpha,\beta}_{\omega,V}\right), \quad K^{\alpha,\beta}_{\omega,V}(u_0) \geq 0, \tag{32}$$

then the solution u to Eq. (1) with Eq. (3) exists globally in time.

1.1 Main theorem

In the previous paper, the authors handled the solution u to Eq. (1) with initial data u_0 satisfying $S_{\omega,V}(u_0) < m_{\omega,V}$ and $K_V(u_0) \geq 0$, where $m_{\omega,V}$ denotes $n_{\omega,V}$ or $r_{\omega,V}$. We note that $m_{\omega,V}$ is $m^{\alpha,\beta}_{\omega,V}$ with $(\alpha, \beta) = (d, 2)$ and $m^{\alpha,\beta}_{\omega,V}$ is independent of (α, β). In this chapter, we are interested in the solutions to Eq. (1) with initial data satisfying $S_{\omega,V}(u_0) < m_{\omega,V}$ and $K_V(u_0) < 0$. Our main theorem is the following:

Theorem 6 Let $d \geq 3$ and $1 + \frac{4}{d} < p < 1 + \frac{4}{d-2}$.

- (Non-radial case) Let $u_0 \in H^1(\mathbb{R}^d)$ and $Q_{\omega,0} \in \mathcal{G}_{\omega,0}$. Suppose that V satisfy "(A1) or (A4) with $|\mathfrak{a}| = 0$," (A2) with $|\mathfrak{a}| = 1$, and (A6). We also assume that there exists $\omega > 0$ such that

$$S_{\omega,V}(u_0) < S_{\omega,0}(Q_{\omega,0}) \ (= n_{\omega,V}), \quad K_V(u_0) < 0. \tag{33}$$

  Then, the solution u to Eq. (1) with Eq. (3) blows up or grows up. Moreover, u blows up under the additional assumption $xu_0 \in L^2(\mathbb{R}^d)$.

- (Radial case) Let $u_0 \in H^1_{\text{rad}}(\mathbb{R}^d)$ and $Q_{\omega,V} \in \mathcal{G}_{\omega,V,\text{rad}}$. Suppose that V is radially symmetric and satisfies "(A1) or (A5) with $|\alpha| = 0$," (A3) with $|\alpha| = 1, 2$, (A7), and $3x \cdot \nabla V + x\nabla^2 V x^T \le 0$. We also assume that there exists $\omega > 0$ satisfying $\omega \ge \omega_0$ such that

$$S_{\omega,V}(u_0) < S_{\omega,V}(Q_{\omega,V}) \ \ (= r_{\omega,V}), \quad K_V(u_0) < 0. \tag{34}$$

Then, the solution u to Eq. (1) with Eq. (3) blows up.

Remark 3 Let V be a potential in Theorem 6. Combining Theorems 5 and 6, we complete bounded and unbounded dichotomy of $\{u_0 \in H^1(\mathbb{R}^d) : S_{\omega,V}(u_0) < S_{\omega,0}(Q_{\omega,0})\}$ and global existence and blow-up dichotomy of $\{u_0 \in H^1_{\text{rad}}(\mathbb{R}^d) : S_{\omega,V}(u_0) < S_{\omega,V}(Q_{\omega,V})\}$ by using sign of the virial functional of initial data.

Remark 4 The following potential satisfies all of conditions in Theorem 6:

$$V(x) = \frac{\gamma\{\log(1+|x|)\}^{\theta}}{|x|^{\mu}}, \quad (\gamma > 0, 0 \le \theta \le \mu < 2, \mu > 0). \tag{35}$$

Theorem 6 with the potential Eq. (35) having $\theta = 0$ was considered in the previous paper [19] by the authors. As the other example, we put

$$V(x) := \frac{\gamma}{\langle x \rangle^{\mu}}, \quad (\gamma > 0, 0 < \mu < 2), \tag{36}$$

where $\langle \cdot \rangle$ is called the Japanese bracket and is defined as $\left(1 + |\cdot|^2\right)^{\frac{1}{2}}$.

1.2 Organization of the paper

The organization of the rest of this chapter is as follows. In Section 2, we collect some notations and tools used throughout this chapter. In Section 3, we prove non-radial case in Theorem 6 by using an argument in [13]. In Section 4, we show radial case in Theorem 6 by using an argument in [33].

2. Preliminaries

In this section, we define some notations and collect some tools, which are used throughout this chapter.

2.1 Notation and definition

For $1 \le p \le \infty$, $L^p = L^p(\mathbb{R}^d)$ denotes the usual Lebesgue space. For a Banach space X, we use $L^q(I; X)$ to denote the Banach space of functions $f : I \times \mathbb{R}^d \to \mathbb{C}$ whose norm is

$$\|f\|_{L^q(I;X)} := \| \|f(t)\|_X \|_{L^q(I)} < \infty. \tag{37}$$

We extend our notation as follows: If a time interval is not specified, then the t-norm is evaluated over $(-\infty, \infty)$. To indicate a restriction to a time subinterval $I \subset (-\infty, \infty)$, we will write as $L^q(I)$. $H^s(\mathbb{R}^d)$ and $\dot{H}^s(\mathbb{R}^d)$ are the usual Sobolev spaces, whose norms $\|f\|_{H^s} := \|(1-\Delta)^{\frac{s}{2}} f\|_{L^2}$ and $\|f\|_{\dot{H}^s} := \|(-\Delta)^{\frac{s}{2}} f\|_{L^2}$ respectively. We also define the Sobolev spaces $H_V^s(\mathbb{R}^d)$ and $\dot{H}_V^s(\mathbb{R}^d)$ with the potential V via norms $\|f\|_{H_V^s} := \|(1-\Delta_V)^{\frac{s}{2}} f\|_{L^2}$ and $\|f\|_{\dot{H}_V^s} := \|(-\Delta_V)^{\frac{s}{2}} f\|_{L^2}$ respectively.

2.2 Some tools

Proposition 2 Let $p \geq 1$. For $f \in H^1_{\text{rad}}(\mathbb{R}^d)$, we have

$$\|f\|^{p+1}_{L^{p+1}(R \leq |x|)} \leq \frac{C}{R^{\frac{(d-1)(p-1)}{2}}} \|f\|^{\frac{p+3}{2}}_{L^2(R \leq |x|)} \|f\|^{\frac{p-1}{2}}_{\dot{H}^1(R \leq |x|)} \tag{38}$$

for any $R > 0$, where the implicit constant C is independent of R and f.
To state the next proposition, we define two functions:

$$\mathcal{X}_R := R^2 \mathcal{X}\left(\frac{|x|}{R}\right), \tag{39}$$

where $\mathcal{X} : [0, \infty) \to [0, \infty)$ (forms)

$$\mathcal{X}(r) := \begin{cases} r^2 & (0 \leq r \leq 1), \\ smooth & (1 \leq r \leq 3), \\ 0 & (3 \leq r) \end{cases} \tag{40}$$

and satisfies $\mathcal{X}''(r) \leq 2$.

$$\mathcal{Y}_R(x) := \mathcal{Y}\left(\frac{|x|}{R}\right), \tag{41}$$

where $\mathcal{Y} : [0, \infty) \to [0, \infty)$ (forms)

$$\mathcal{Y}(r) := \begin{cases} 0 & \left(0 \leq r \leq \frac{1}{2}\right), \\ smooth & \left(\frac{1}{2} \leq r \leq 1\right), \\ 1 & (1 \leq r) \end{cases} \tag{42}$$

and satisfies $0 \leq \mathcal{Y}'(r) \leq 3$.
Proposition 3 (Localized virial identity, [3]) Let w be $\mathcal{X}_R$ or $\mathcal{Y}_R$ defined as Eqs. (39) and (41) respectively. For the solution u to Eq. (1), we define

$$I_w(t) := \int_{\mathbb{R}^d} w(x)|u(t,x)|^2 dx. \tag{43}$$

Then, we have

$$\begin{aligned}I_{w'}(t) &= 2\mathrm{Im}\int_{\mathbb{R}^d}\frac{x\cdot\nabla u}{|x|}\overline{u}w'dx,\\ I_{w''}(t) &= \int_{\mathbb{R}^d}F_1|x\cdot\nabla u|^2dx + 4\int_{\mathbb{R}^d}\frac{w'}{|x|}|\nabla u|^2dx - \int_{\mathbb{R}^d}F_2|u|^{p+1}dx\\ &\quad-\int_{\mathbb{R}^d}F_3|u|^2dx - 2\int_{\mathbb{R}^d}\frac{w'}{|x|}(x\cdot\nabla V)|u|^2dx.\end{aligned} \tag{44}$$

where

$$\begin{aligned}&F_1(w,|x|) := 4\left(\frac{w''}{|x|^2}-\frac{w'}{|x|^3}\right), \qquad F_2(w,|x|) := \frac{2(p-1)}{p+1}\left(w''+\frac{d-1}{|x|}w'\right),\\ &F_3(w,|x|) := w^{(4)} + \frac{2(d-1)}{|x|}w^{(3)} + \frac{(d-1)(d-3)}{|x|^2}w'' + \frac{(d-1)(3-d)}{|x|^3}w'.\end{aligned} \tag{45}$$

3. Non-radial case of main theorem

In this section, we prove (Non-radial case) for Theorem 6. First, we recall rewriting of $n_{\omega,V}$, which is given in [40].

Lemma 1 Let $d\geq 3$, $1+\frac{4}{d}<p<1+\frac{4}{d-2}$, and $Q_{\omega,0}\in\mathcal{G}_{\omega,0}$. Assume that V satisfies (A2) with $|\alpha|\leq 1$ and (A6). Then,

$$S_{\omega,0}(Q_{\omega,0}) = n_{\omega,V} = \inf\{T_{\omega,V}(f) : f\in H^1(\mathbb{R}^d)\backslash\{0\},\ K_V(f)\leq 0\} \tag{46}$$

holds, where the functional $T_{\omega,V}$ is defined as

$$T_{\omega,V}(f) := S_{\omega,V}(f) - \frac{1}{4}K_V(f). \tag{47}$$

Next, we give uniform estimate of the virial functional K_V.

Lemma 2 Under the all assumptions of (Non-radial) in Theorem 6, there exists $\delta>0$ such that

$$\sup_{t\in(T_{\min},T_{\max})} K_V(u(t)) \leq -\delta < 0. \tag{48}$$

Proof: Let $\delta := 4\{S_{\omega,V}(Q_{\omega,V}) - S_{\omega,V}(u_0)\} > 0$. Applying Lemma 1, we have

$$\begin{aligned}S_{\omega,V}(Q_{\omega,V}) &\leq T_{\omega,V}(u(t)) = S_{\omega,V}(u_0) - \frac{1}{4}K_V(u(t))\\ &= S_{\omega,V}(Q_{\omega,V}) - \frac{1}{4}\delta - \frac{1}{4}K_V(u(t)),\end{aligned} \tag{49}$$

which implies the desired result.

The blow-up result with $xu_0\in L^2(\mathbb{R}^d)$ of (Non-radial case) in Theorem 1.1 follows immediately from Lemma 2.

Proof of blow-up part in (Non-radial case) for Theorem 6: We assume that the solution u exists globally in time for contradiction. When $xu_0\in L^2(\mathbb{R}^d)$, we have Eq. (22). Combining Eq. (22) and Lemma 2, there exists $\delta>0$ such that

$$\frac{d^2}{dt^2}\|xu(t)\|_{L^2}^2 = 4K_V(u(t)) < -4\delta < 0 \tag{50}$$

for any $t \in \mathbb{R}$. Therefore, we obtain $\|xu(t)\|_{L^2}^2 < 0$ if $|t|$ is sufficiently large. However, this is contradiction.

We consider Lemmas 3 and 4 to prove blow-up or grow-up part in (Non-radial case) for Theorem 6.

Lemma 3 Let $d \geq 3$ and $1 + \frac{4}{d} < p < 1 + \frac{4}{d-2}$. We assume that $u \in C([0, \infty); H^1)$ be a solution to Eq. (1) satisfying $C_0 := \sup_{t \in [0,\infty)} \|u(t)\|_{\dot{H}_x^1} < \infty$. Then, it follows that

$$\|u(t)\|_{L^2(|x| \geq R)}^2 \leq o_R(1) + \eta \tag{51}$$

for any $\eta > 0$, $R > 0$, and $t \in \left[0, \frac{\eta R}{6C_0\|u\|_{L_x^2}}\right]$, where $o_R(1)$ goes to zero as $R \to \infty$ and is independent of t.

Proof: We consider $I_{\mathcal{Y}_R}$ given in Eq. (43). Using Proposition 3,

$$\begin{aligned} I(t) &= I(0) + \int_0^t I'(s)ds \leq I(0) + \int_0^t |I'(s)|\, ds \\ &\leq I(0) + \frac{2t}{R}\|\mathcal{Y}'\|_{L^\infty} \sup_{t \in [0,\infty)} \|u(t)\|_{\dot{H}_x^1}\|u\|_{L_x^2} \leq I(0) + \frac{6C_0\|u\|_{L_x^2} t}{R} \end{aligned} \tag{52}$$

for any $t \in [0, \infty)$. By the definition of $\mathcal{Y}_R$, we have

$$I(0) = \int_{\mathbb{R}^d} \mathcal{Y}_R(x)|u_0(x)|^2 dx \leq \|u_0\|_{L^2\left(|x| \geq \frac{R}{2}\right)}^2 = o_R(1) \tag{53}$$

and hence, we obtain

$$\|u(t)\|_{L^2(|x| \geq R)}^2 \leq I(t) \leq o_R(1) + \eta. \tag{54}$$

Lemma 4 Let $d \geq 3$ and $1 + \frac{4}{d} < p < 1 + \frac{4}{d-2}$. Let $u \in C([0, \infty); H^1(\mathbb{R}^d))$ be a solution to Eq. (1). Then, for $q \in (p + 1, 2^*)$, there exist constants $C = C(q, \|u_0\|_{L^2}, C_0) > 0$ and $\theta_q > 0$ such that the estimate

$$I_{\mathcal{X}_R}{}'(t) \leq 4K_V(u(t)) + C\|u(t)\|_{L^2(R \leq |x|)}^{(p+1)\theta_q} + \frac{C}{R^2} \tag{55}$$

holds for any $R > 0$ and $t \in [0, \infty)$, where $\theta_q := \frac{2\{q-(p+1)\}}{(p+1)(q-2)} \in \left(0, \frac{2}{p+1}\right)$, C_0 is given in Lemma 3, and $I_{\mathcal{X}_R}$ is defined as Eq. (43).

Proof: Using Proposition 3, we have

$$I''_{\mathcal{X}_R}(t) = 4K_V(u(t)) + \mathcal{R}_1 + \mathcal{R}_2 + \mathcal{R}_3 + \mathcal{R}_4, \tag{56}$$

where $\mathcal{R}_k = \mathcal{R}_k(t)$ $(k = 1,2,3,4)$ are defined as

$$\begin{aligned} \mathcal{R}_1 &:= 4\int_{\mathbb{R}^d} \left\{\frac{1}{|x|^2}\mathcal{X}''\left(\frac{r}{R}\right) - \frac{R}{|x|^3}\mathcal{X}'\left(\frac{|x|}{R}\right)\right\}|x \cdot \nabla u|^2 dx \\ &\quad + 4\int_{\mathbb{R}^d} \left\{\frac{R}{|x|}\mathcal{X}'\left(\frac{|x|}{R}\right) - 2\right\}|\nabla u(t,x)|^2 dx, \end{aligned} \tag{57}$$

$$\mathcal{R}_2 := -\frac{2(p-1)}{p+1}\int_{\mathbb{R}^d}\left\{\mathcal{X}''\left(\frac{|x|}{R}\right)+\frac{(d-1)R}{|x|}\mathcal{X}'\left(\frac{|x|}{R}\right)-2d\right\}|u(t,x)|^{p+1}dx, \tag{58}$$

$$\begin{aligned}\mathcal{R}_3 := &-\int_{\mathbb{R}^d}\left\{\frac{1}{R^2}\mathcal{X}^{(4)}\left(\frac{|x|}{R}\right)+\frac{2(d-1)}{R|x|}\mathcal{X}^{(3)}\left(\frac{|x|}{R}\right)+\frac{(d-1)(d-3)}{|x|^2}\mathcal{X}\prime'\left(\frac{|x|}{R}\right)\right.\\ &\left.+\frac{(d-1)(3-d)R}{|x|^3}\mathcal{X}'\left(\frac{|x|}{R}\right)\right\}|u(t,x)|^2dx,\end{aligned} \tag{59}$$

$$\mathcal{R}_4 := 2\int_{R\le|x|}\left\{2-\frac{R}{|x|}\mathcal{X}'\left(\frac{|x|}{R}\right)\right\}(x\cdot\nabla V)|u(t,x)|^2dx. \tag{60}$$

We set

$$\Omega := \left\{x\in\mathbb{R}^d : \frac{1}{|x|^2}\mathcal{X}''\left(\frac{|x|}{R}\right)-\frac{R}{|x|^3}\mathcal{X}'\left(\frac{|x|}{R}\right)\le 0\right\}. \tag{61}$$

By the inequality $\mathcal{X}'\left(\frac{|x|}{R}\right)\le\frac{2|x|}{R}$, we have

$$\mathcal{R}_1 \le 4\int_{\Omega^c}\left\{\mathcal{X}''\left(\frac{r}{R}\right)-2\right\}|\nabla u(t,x)|^2dx \le 0, \tag{62}$$

where Ω^c denotes a complement of Ω.

Next, we estimate $\mathcal{R}_2$. Applying Hölder's inequality and Sobolev's embedding, we have

$$\begin{aligned}\mathcal{R}_2 &\le C\|u(t)\|_{L^{p+1}(R\le|x|)}^{p+1} \le C\|u(t)\|_{L^q(R\le|x|)}^{(p+1)(1-\theta_q)}\|u(t)\|_{L^2(R\le|x|)}^{(p+1)\theta_q}\\ &\le C\|u(t)\|_{H^1}^{(p+1)(1-\theta_q)}\|u(t)\|_{L^2(R\le|x|)}^{(p+1)\theta_q} \le C\|u(t)\|_{L^2(R\le|x|)}^{(p+1)\theta_q}.\end{aligned} \tag{63}$$

Next, we estimate $\mathcal{R}_3$.

$$\mathcal{R}_3 \le \frac{C}{R^2}\|u(t)\|_{L^2(R\le|x|)}^2 \le \frac{C}{R^2}. \tag{64}$$

Finally, $\mathcal{R}_4$ is estimated as $\mathcal{R}_4\le 0$ by $\mathcal{X}'\left(\frac{|x|}{R}\right)\le\frac{2|x|}{R}$ and $x\cdot\nabla V\le 0$, which completes the proof of the lemma.

Proof of blow-up or grow-up part in (Non-radial case) for Theorem 6. We assume that

$$T_{\max}=\infty \quad \text{and} \quad \sup_{t\in[0,\infty)}\|u(t)\|_{\dot{H}_x^1}<\infty \tag{65}$$

for contradiction. By Lemmas 2, 3, and 4, there exists $\delta>0$ such that

$$I''_{\mathcal{X}_R}(s) \le -4\delta + C\|u(s)\|_{L_x^2(R\le|x|)}^{(p+1)\theta_q}+\frac{C}{R^2} \le -4\delta + C\eta^{\frac{(p+1)\theta_q}{2}}+o_R(1) \tag{66}$$

for any $\eta>0$, $R>0$, and $s\in\left[0,\frac{\eta R}{6C_0\|u_0\|_{L^2}}\right]$. We take $\eta=\eta_0>0$ sufficiently small such as

$$C\eta_0^{\frac{(p+1)\theta_q}{2}} \leq 2\delta. \tag{67}$$

and set

$$T = T(R) := \alpha_0 R := \frac{\eta_0 R}{6C_0\|u_0\|_{L^2}}. \tag{68}$$

Applying Eq. (67), integrating Eq. (66) over $s \in [0, t]$, and integrating over $t \in [0, T]$, we have

$$\begin{aligned} I_{\chi_R}(T) &\leq I_{\chi_R}(0) + I'_{\chi_R}(0)T + \frac{1}{2}(-2\delta + o_R(1))T^2 \\ &= I_{\chi_R}(0) + I'_{\chi_R}(0)\alpha_0 R + \frac{1}{2}(-2\delta + o_R(1))\alpha_0^2 R^2. \end{aligned} \tag{69}$$

Here, we can see

$$I_{\chi_R}(0) = o_R(1)R^2 \quad \text{and} \quad I'_{\chi_R}(0) = o_R(1)R. \tag{70}$$

Indeed, we get

$$I_{\chi_R}(0) \leq R\|u_0\|^2_{L^2(|x| \leq \sqrt{R})} + cR^2\|u_0\|_{L^2(\sqrt{R} \leq |x|)} = o_R(1)R^2, \tag{71}$$

and

$$I'_{\chi_R}(0) \leq 4\sqrt{R}\|u_0\|_{\dot{H}^1}\|u_0\|_{L^2(|x| \leq \sqrt{R})} + cR\|u_0\|_{\dot{H}^1}\|u_0\|_{L^2(\sqrt{R} \leq |x|)} = o_R(1)R. \tag{72}$$

Combining Eqs. (69) and (70), we get

$$I_{\chi_R}(T) \leq \left(o_R(1) - \delta\alpha_0^2\right)R^2. \tag{73}$$

We take $R > 0$ such as $o_R(1) - \delta\alpha_0^2 < 0$. However, this contradicts $I_{\chi_R}(T) \geq 0$.

4. Radial case of main theorem

In this section, we prove (Radial case) for Theorem 6. First, we introduce another characterization of $r_{\omega,V}$.

Lemma 5 Let $d \geq 3$, $1 + \frac{4}{d} < p < 1 + \frac{4}{d-2}$, and $Q_{\omega,V} \in \mathcal{G}_{\omega,V,\text{rad}}$. Assume that V is radially symmetric and satisfies (A3) with $|\alpha| \leq 2$, (A7), and $3x \cdot \nabla V + x\nabla^2 V x^T \leq 0$. Then,

$$S_{\omega,V}(Q_{\omega,V}) = r_{\omega,V} = \inf\{U_{\omega,V}(f) : f \in H^1_{\text{rad}}(\mathbb{R}^d)\backslash\{0\},\ K_V(f) \leq 0\} \tag{74}$$

holds, where the functional $U_{\omega,V}$ is defined as

$$U_{\omega,V}(f) := S_{\omega,V}(f) - \frac{1}{d(p-1)}K_V(f). \tag{75}$$

Proof: The lemma follows from proof of Lemma 1 (see [40], Lemma 4.3) combined $2\omega + 2V + x \cdot \nabla V \geq 0$.

Proof of (Radial case) for Theorem 6. Assume that the solution u to Eq. (1) exists globally in time for contradiction. We consider $I_{\mathcal{X}_R}$ again and recall

$$I''_{\mathcal{X}_R}(t) = 4K_V(u(t)) + \mathcal{R}_1 + \mathcal{R}_2 + \mathcal{R}_3 + \mathcal{R}_4, \tag{76}$$

where $\mathcal{R}_k$ $(1 \le k \le 4)$ are defined as Eqs. (57) $\sim$ (60). We use same estimates with proof of blow-up or grow-up for $\mathcal{R}_1$, $\mathcal{R}_3$, and $\mathcal{R}_4$. Applying Proposition 2 and the Young's inequality, we have

$$\begin{aligned}
\mathcal{R}_2 &\le \frac{C}{R^{\frac{(d-1)(p-1)}{2}}} \|u(t)\|_{L^2(R \le |x|)}^{\frac{p+3}{2}} \|u(t)\|_{\dot{H}^1(R \le |x|)}^{\frac{p-1}{2}} \\
&\le \frac{C}{R^{\frac{2(d-1)(p-1)}{5-p}} \varepsilon^{\frac{4}{5-p}}} \|u\|_{L^2}^{\frac{2(p+3)}{5-p}} + 2\{d(p-1) - 4\}\varepsilon \|u\|_{\dot{H}^1}^2 \\
&\le \frac{C}{R^{\frac{2(d-1)(p-1)}{5-p}} \varepsilon^{\frac{4}{5-p}}} \|u\|_{L^2}^{\frac{2(p+3)}{5-p}} + 4d(p-1)\varepsilon U_{\omega,V}(u)
\end{aligned} \tag{77}$$

for each positive $\varepsilon > 0$, which is chosen later. Collecting these estimates, we have

$$\begin{aligned}
&I''_{\mathcal{X}_R}(t) \\
&\le 4K_V(u) + 4d(p-1)\varepsilon U_{\omega,V}(u) + \frac{C}{R^{\frac{2(d-1)(p-1)}{5-p}} \varepsilon^{\frac{4}{5-p}}} + \frac{C}{R^2} \\
&= 4d(p-1)\{S_{\omega,V}(u) - U_{\omega,V}(u)\} + 4d(p-1)\varepsilon U_{\omega,V}(u) + \frac{C}{R^{\frac{2(d-1)(p-1)}{5-p}} \varepsilon^{\frac{4}{5-p}}} + \frac{C}{R^2} \\
&< 4d(p-1)(1-\delta)S_{\omega,V}(Q_{\omega,V}) + 4d(p-1)(\varepsilon - 1)U_{\omega,V}(u) + \frac{C}{R^{\frac{2(d-1)(p-1)}{5-p}} \varepsilon^{\frac{4}{5-p}}} + \frac{C}{R^2} \\
&\le 4d(p-1)(\varepsilon - \delta)S_{\omega,V}(Q_{\omega,V}) + \frac{C}{R^{\frac{2(d-1)(p-1)}{5-p}} \varepsilon^{\frac{4}{5-p}}} + \frac{C}{R^2},
\end{aligned} \tag{78}$$

where the second inequality is used $S_{\omega,V}(u) < (1-\delta)S_{\omega,V}(Q_{\omega,V})$ for some $\delta \in (0,1)$ and the third inequality is used $S_{\omega,V}(Q_{\omega,V}) \le U_{\omega,V}$ (see Lemma 5). Taking $\varepsilon \in (0, \delta)$ and sufficiently large $R > 0$, there exists $\eta > 0$ such that $I''_{\mathcal{X}_R}(t) < -\eta < 0$ for each $t \in \mathbb{R}$. However, this inequality implies that if $|t|$ is sufficiently large, then . This is contradiction and hence, we complete the proof.

5. Conclusions

In this chapter, our main result is Theorem 6. Combining the main result and a previous result (Theorem 5), we can classify time behavior of solutions to Eq. (1) with initial data below the ground state in the sense of their action $S_{\omega,V}$ by using sign of the virial functional for the initial data. More precisely, for the solution $u(t)$ with $S_{\omega,V}(u_0) < S_{\omega,0}(Q_{\omega,0})$, if $K_V(u_0) \ge 0$ then u is bounded in $H^1(\mathbb{R}^d)$ and if $K_V(u_0) < 0$ then u is unbounded in $H^1(\mathbb{R}^d)$. In addition, for the radial solution $u(t)$ with $S_{\omega,V}(u_0) < S_{\omega,V}(Q_{\omega,V})$, if $K_V(u_0) \ge 0$, then u exists globally in time and if $K_V(u_0) < 0$ then u blows up.

Acknowledgements

M.H. is supported by JSPS KAKENHI Grant Number JP22J00787. M.I. is supported by JSPS KAKENHI Grant Number JP19K14581 and JST CREST Grant Number JPMJCR1913.

Conflict of interest

The authors declare no conflict of interest.

Author details

Masaru Hamano[1*†] and Masahiro Ikeda[2,3†]

1 Faculty of Science and Engineering, Waseda University, Tokyo, Japan

2 RIKEN Center for Advanced Intelligence Project, Tokyo, Japan

3 Department of Mathematics, Faculty of Science and Technology, Keio University, Yokohama, Japan

*Address all correspondence to: m.hamano3@kurenai.waseda.jp

† These authors contributed equally.

References

[1] Rose HA, Weinstein MI. On the bound states of the nonlinear Schrödinger equation with a linear potential. Physica D. 1988;**30**(1–2): 207-218

[2] Cazenave T. Semilinear Schrödinger Equations, Courant Lecture Notes in Mathematics. Vol. **10**. New York; Providence, RI: New York University, Courant Institute of Mathematical Sciences; American Mathematical Society; 2003. p. xiv+323. MR2002047

[3] Hong Y. Scattering for a nonlinear Schrödinger equation with a potential. Communications on Pure and Applied Analysis. 2016;**15**(5):1571-1601. MR3538870

[4] Hamano M, Ikeda M. Global dynamics below the ground state for the focusing Schrödinger equation with a potential. Journal of Evolution Equations. 2020;**20**(3):1131-1172. MR4142248

[5] Akahori T, Nawa H. Blowup and scattering problems for the nonlinear Schrödinger equations. Kyoto Journal of Mathematics. 2013;**53**(3):629-672. MR3102564

[6] Arora AK, Dodson B, Murphy J. Scattering below the ground state for the 2d radial nonlinear Schrödinger equation. Proceedings of the American Mathematical Society. 2020;**148**(4): 1653-1663. MR4069202

[7] Dinh VD. On nonlinear Schrödinger equations with attractive inverse-power potentials. Topological Methods in Nonlinear Analysis. 2021;**57**(2):489-523. MR4359723

[8] Dinh VD. On nonlinear Schrödinger equations with repulsive inverse-power potentials. Acta Applicandae Mathematicae. 2021;**171**:52. Paper No. 14, MR4198524

[9] Dodson B. Global well-posedness and scattering for the mass critical nonlinear Schrödinger equation with mass below the mass of the ground state. Advances in Mathematics. 2015;**285**:1589-1618. MR3406535

[10] Dodson B. Global well-posedness and scattering for the focusing, cubic Schrödinger equation in dimension $d = 4$. Annales Scientifiques de l'École Normale Supérieure. 2019;**52**(1): 139-180. MR3940908

[11] Dodson B, Murphy J. A new proof of scattering below the ground state for the 3D radial focusing cubic NLS. Proceedings of the American Mathematical Society. 2017;**145**(11): 4859-4867. MR3692001

[12] Dodson B, Murphy J. A new proof of scattering below the ground state for the non-radial focusing NLS. Mathematical Research Letters. 2018;**25**(6):1805-1825. MR3934845

[13] Du D, Wu Y, Zhang K. On blow-up criterion for the nonlinear Schrödinger equation. Discrete and Continuous Dynamical Systems. 2016;**36**(7): 3639-3650. MR3485846

[14] Duyckaerts T, Holmer J, Roudenko S. Scattering for the non-radial 3D cubic nonlinear Schrödinger equation. Mathematical Research Letters. 2008;**15**(6):1233-1250. MR2470397

[15] Fang D, Xie J, Cazenave T. Scattering for the focusing energy-subcritical nonlinear Schrödinger equation. Science

China Mathematics. 2011;**54**(10): 2037-2062. MR2838120

[16] Farah LG, Pastor A. Scattering for a 3D coupled nonlinear Schrödinger system. Journal of Mathematical Physics. 2017;**58**(7):33, 071502. MR3671163

[17] Glassey RT. On the blowing up of solutions to the Cauchy problem for nonlinear Schrödinger equations. Journal of Mathematical Physics. 1977;**18**(9): 1794-1797. MR0460850

[18] Hamano M. Global dynamics below the ground state for the quadratic Schödinger system in 5d. Preprint, arXiv: 1805.12245

[19] Hamano M, Ikeda M. Equivalence of conditions on initial data below the ground state to NLS with a repulsive inverse power potential. Journal of Mathematical Physics. 2022;**63**(3):16. Paper No. 031509, MR4393612

[20] Hamano M, Ikeda M. Scattering solutions to nonlinear Schrödinger equation with a long range potential. Journal of Mathematical Analysis and Applications. 2023;**528**(1). Paper No. 127468. MR4602980

[21] Hamano M, Ikeda M, Inui T, Shimizu I. Global dynamics below a threshold for the nonlinear Schrödinger equations with the Kirchhoff boundary and the repulsive Dirac delta boundary on a star graph. Preprint, arXiv: 2212.06411

[22] Hamano M, Inui T, Nishimura K. Scattering for the quadratic nonlinear Schrödinger system in $\mathbb{R}^5$ without mass-resonance condition. Fako de l'Funkcialaj Ekvacioj Japana Matematika Societo. 2021;**64**(3):261-291. MR4360610

[23] Holmer J, Roudeko S. A sharp condition for scattering of the radial 3D cubic nonlinear Schrödinger equation. Communications in Mathematical Physics. 2008;**282**(2):435-467. MR2421484

[24] Ibrahim S, Masmoudi N, Nakanishi K. Scattering threshold for the focusing nonlinear Klein-Gordon equation. Analysis of PDEs. 2011;**4**(3): 405-460. MR2872122

[25] Ikeda M, Inui T. Global dynamics below the standing waves for the focusing semilinear Schrödinger equation with a repulsive Dirac delta potential. Analysis of PDEs. 2017;**10**(2): 481-512. MR3619878

[26] Inui T, Kishimoto N, Nishimura K. Scattering for a mass critical NLS system below the ground state with and without mass-resonance condition. Discrete and Continuous Dynamical Systems. 2019; **39**(11):6299-6353. MR4026982

[27] Inui T, Kishimoto N, Nishimura K. Blow-up of the radially symmetric solutions for the quadratic nonlinear Schrödinger system without mass-resonance. Nonlinear Analysis. 2020; **198**:10, 111895. MR4090442

[28] Kenig CE, Merle F. Global well-posedness, scattering and blow-up for the energy-critical, focusing, non-linear Schrödinger equation in the radial case. Inventiones Mathematicae. 2006;**166**(3): 645-675. MR2257393

[29] Kenig CE, Merle F. Global well-posedness, scattering and blow-up for the energy-critical focusing non-linear wave equation. Acta Mathematica. 2008; **201**(2):147-212. MR2461508

[30] Killip R, Murphy J, Visan M, Zheng J. The focusing cubic NLS with inverse-square potential in three space

dimensions. Differential and Integral Equations. 2017;**30**(3–4):161-206. MR3611498

[31] Killip R, Visan M. The focusing energy-critical nonlinear Schrödinger equation in dimensions five and higher. American Journal of Mathematics. 2010; **132**(2):361-424. MR2654778

[32] Lu J, Miao C, Murphy J. Scattering in H^1 for the intercritical NLS with an inverse-square potential. Journal of Differential Equations. 2018;**264**(5): 3174-3211. MR3741387

[33] Ogawa T, Tsutsumi Y. Blow-up of H^1 solution for the nonlinear Schrödinger equation. Journal of Differential Equations. 1991;**92**(2):317-330. MR1120908

[34] Wang H, Yang Q. Scattering for the 5D quadratic NLS system without mass-resonance. Journal of Mathematical Physics. 2019;**60**(12):23, 121508. MR4043361

[35] Xu G. Dynamics of some coupled nonlinear Schrödinger systems in $\mathbb{R}^3$. Mathematicsl Methods in the Applied Sciences. 2014;**37**(17):2746-2771. MR3271121

[36] Xu C. Scattering for the non-radial focusing inhomogeneous nonlinear Schrödinger-Choquard equation. Preprint, arXiv: 2104.09756

[37] Zhang J, Zheng J. Scattering theory for nonlinear Schrödinger equations with inverse-square potential. Journal of Functional Analysis. 2014;**267**(8): 2907-2932. MR3255478

[38] Zheng J. Focusing NLS with inverse square potential. Journal of Mathematical Physics. 2018;**59**(11):14, 111502. MR3872306

[39] Mizutani H. Wave operators on Sobolev spaces. Proceedings of the American Mathematical Society. 2020; **148**(4):1645-1652. MR4069201

[40] Hamano M, Ikeda M. Characterization of the ground state to the intercritical NLS with a linear potential by the virial functional. In: Advances in Harmonic Analysis and Partial Differential Equations, Trends Math. Cham: Birkhäuser/Springer; 2020. pp. 279-307. MR4174752

[41] Hamano M, Ikeda M. Global well-posedness below the ground state for the nonlinear Schrödinger equation with a linear potential. Proceedings of the American Mathematical Society. 2020; **148**(12):5193-5207. MR4163832